Zhiyun Lin

Distributed Control and Analysis of Coupled Cell Systems

Zhiyun Lin

Distributed Control and Analysis of Coupled Cell Systems

VDM Verlag Dr. Müller

Imprint

Bibliographic information by the German National Library: The German National Library lists this publication at the German National Bibliography; detailed bibliographic information is available on the Internet at http://dnb.d-nb.de.

Cover image: www.purestockx.com

Publisher:
VDM Verlag Dr. Müller Aktiengesellschaft & Co. KG, Dudweiler Landstr. 125 a, 66123 Saarbrücken, Germany,
Phone +49 681 9100-698, Fax +49 681 9100-988,
Email: info@vdm-verlag.de

Produced in USA and UK by:
Lightning Source Inc., La Vergne, Tennessee, USA
Lightning Source UK Ltd., Milton Keynes, UK
BookSurge LLC, 5341 Dorchester Road, Suite 16, North Charleston, SC 29418, USA

ISBN: 978-3-639-00784-8

To Ivy

for love and fun

Preface

A system composed of many simple entities, each obeying the same simple rules of interaction, often displays complex collective behaviors. Such emergent behaviors have been observed in a multitude of physical, biological, and social systems, for example the swirling flight of bird flocks or the purposeful social activities of insect colonies. Global system behavior emerges from local agent interactions despite the absence of centralized coordination and global information exchange. Coordination among interacting agents is not only a recognized phenomenon in nature, but also an engineering problem related to the control aspect of mobile robots, sensor networks, and satellite clusters (e.g., rendezvous, deployment, formation control, search and rescue).

We regard a system composed of many agents interacting each other as a *coupled cell system* in continuous time and initiate a systematic inquiry from structural point of view. The goal is to develop a framework and methodology for the analysis of collective behaviors on the one hand and the synthesis of bio-inspired distributed control for engineered systems on the other hand.

Stability and agreement properties, that capture crucial components of many desirable features of collective behaviors, such as swarming, flocking, rendezvous, and synchronization, are investigated in detail. Rigorous analysis is provided under a unified mathematical model of switched interconnected system. In the study, the interconnection topologies are represented schematically by directed graphs called *interaction graphs*. Both fixed topology and dynamic topology— subsystems become disconnected from each other and may again become connected to each other for various natural or technological reasons, are treated. Under certain cooperative assumptions on individual dynamics, it is shown that some technical connectivity properties of the interaction graph lead to stability and agreement of the overall system.

In control aspect, we restrict to the coordinated control of autonomous mobile robots. Autonomous means that the robot has only sensory input—no outside direct commands. The aim is to control a group of robots connected through a sensor or communication network to accomplish a global task cooperatively. Further, we restrict to distributed control: Each robot has the same local control strategy—no leaders. Within the same framework of coupled cell systems, the problem of distributed coordination and control for multiple mobile robots such as rendezvous (bring all robots to a common meeting point) and attaining a desired formation is solved even when local interactions dynamically change over time.

Toronto
April 2008

Contents

List of Figures

Part I

Introduction

1

Introduction to Coupled Cell Systems

A coupled cell system is a collection of individual, but interacting, dynamical systems. With coupling, the state of certain individual systems (called *cells* or *agents*) affects the time-evolution of other agents. A schematic diagram of such system is shown in Fig. 1.1. The network represents the coupling structure among

Fig. 1.1. Coupled cell systems.

them. It can be a communication link, a sensor network, or other interaction relationships. Moreover, the coupling structure can be static or dynamic when links are established and dropped over time. We ask: How much of the qualitative behaviors observed in macro-level is the product of network architecture and how much is related to the specific dynamics of agents and the way they are coupled?

1.1 Issues in Coupled Cell Systems

Coupled cell systems encompass, or are closely related to, many physical, biological, chemical, and engineering systems. In this section we illustrate some related issues by providing a short literature review. Although they are taken from different fields of study, they are all in the same framework of coupled cell systems and share common problems from the point of view of systems theory.

1.1.1 Emergent Behaviors

Emergent behavior is behavior that looks complex and/or purposeful to the observer but is actually derived spontaneously from fairly simple rules. The lower level entities following the rules have no idea of the bigger picture. They are only aware of themselves and maybe a few of their neighbors. Emergent behavior can be found everywhere in biological systems. Schooling of fishes (Fig. 1.2) is a beautiful demonstration of emergent behavior. Through simple local agent

Fig. 1.2. Schooling of fishes.

interaction, desired cooperative behaviors emerge. Biologists have been working on understanding and modelling of group behaviors for a long time. See for example [26], [104], and references therein (some of which date back to the 1920's). The work by Breder [26] is one of the early efforts to develop mathematical models of the schooling behavior in fish. He suggests a simple model composed of attraction and repulsion components. Recently, in [137], Vicsek et al. propose a simple but compelling discrete-time model of n autonomous agents (i.e., points or particles usually called *self-driven* or *self-propelled particles*), and study the collective behavior due to their interaction, where they assume that particles are moving with constant absolute velocity and at each step each one travels in the average direction of motion of the particles in its neighborhood with some random perturbation. In their paper, Vicsek et al. provide a variety of interesting simulation results which demonstrate that the nearest neighbor rule they are studying can cause all agents to eventually move in the same direction despite the absence of centralized coordination and despite the fact that each agent's set of nearest neighbors changes with time as the system evolves. An earlier model was introduced by Reynolds [111] in 1987. Reynolds wrote a program called *boids* [110] simulating the motion of a flock of birds; they fly as a flock, with a common average heading, and they avoid colliding with each other. Each bird has a local control strategy—there's no leader broadcasting instructions—yet a desirable overall group behavior is achieved. The local strategy of each bird has three components: 1) *separation*, steer to avoid crowding local flock mates; 2) *alignment*, steer towards the average heading of local flock mates; 3) *cohesion*, steer to move toward the average position of local flock mates. Recently, Jadbabaie et al. [58] study the second of these strategies and prove mathematically, under some conditions, that all agents will eventually move in a common direction. Thus, a local strategy produces a global objective. Besides, in [46, 47, 85], stability of synchronous and asynchronous swarms with a fixed communication topology is studied, where stability is used to characterize the cohesiveness of a swarm.

1.1.2 Coupled Nonlinear Oscillators

Mutual synchronization is a common phenomenon in biology. It occurs at different levels, ranging from the small scale of the cardiac pace-maker cells of the SA (Sino-Atrial) and AV (Atrium-Ventricular) nodes in the human heart that synchronously fire and give the pace to the whole muscle, to the coordinated behaviors of crickets that chirp in unison and of fireflies that flash together in some parts of southeast Asia. The dynamics of coupled oscillators is a very broad field of research. Synchronous motion was probably first reported by Huygens ([56], 1673). The subject of synchronization has received huge attention in recent decades. For example, arrays of chaotic systems are studied in [15, 16, 106, 145, 146]. For coupled nonlinear oscillators, a seminal study to understand synchronization is done by Kuramoto in [70]; the work is reviewed in [127, 128]; more recently, the problem has been reinvestigated from the viewpoint of system control in [59, 124]. In addition, synchronization of mechanical systems is dealt with in [98].

Kuramoto model describes the dynamics of a set of n phase oscillators θ_i with natural frequencies ω_i. More details can be found in [59, 70, 127, 128]. The time evolution of the i-th oscillator is given by

$$\dot{\theta}_i = \omega_i + k_i \sum_{j \in \mathcal{N}_i(t)} \sin(\theta_j - \theta_i),$$

where $k_i > 0$ is the coupling strength and $\mathcal{N}_i(t)$ is the set of neighbors of oscillator i at time t. The interaction structure can be general so far, that is, $\mathcal{N}_i(t)$ can be an arbitrary set of other nodes and can be dynamic. The objective is clear here to achieve synchronization of frequencies of n oscillators.

1.1.3 Collective Autonomous Robots

More recently, coordination and control of multi-agent/multi-vehicle systems in the framework of coupled dynamic systems have attracted increasing interests in the field of system control and robotics. Several researchers began investigating distributed algorithms for multi-agent systems in the early 1990s. In [129] a group of simulated robots forms approximations to circles and simple polygons using the scenario that each robot orients itself to the furthest or nearest robot. In [4, 5, 99, 131], some distributed algorithms are proposed with the objective of getting the robots to congregate at a common location (*rendezvous*). These algorithms are extended to various synchronous and asynchronous stop-and-go strategies in [31, 75, 76]. In addition to these modelling and simulation studies, research papers focusing on the detailed mathematical analysis of coupled dynamic systems began to appear. Theoretical development of information consensus and agreement among a network of agents is made in discrete time [58, 94], in continuous time [13, 53, 77, 95, 103, 109, 118], and in a quantized data communication setup [61, 123]. Stabilization of vehicle formations with linear dynamics is studied in [40, 41, 73, 115–117] using potential functions. For vehicles with nonholonomic constraints, achievable equilibrium formations are explored in [64, 82, 89–91, 132, 133, 147, 148]. Other relevant references on formation control are [34–38, 42, 93, 100, 140, 141]. More detailed discussion of these references is postponed to appropriate chapters.

1.1.4 Biochemical Reaction Networks

A biochemical reaction network is a finite set of reactions among a finite set of species. Consider, for example, two reversible reactions among three compounds

C_1, C_2, and C_3, in which C_1 is transformed into C_2, C_2 is transformed into C_3, and vice versa:

$$C_1 \underset{k_2}{\overset{k_1}{\rightleftharpoons}} C_2 \underset{k_4}{\overset{k_3}{\rightleftharpoons}} C_3$$

The constants $k_1 > 0$, $k_2 > 0$ are the forward and reverse rate constants of the reaction $C_1 \rightleftharpoons C_2$; similarly for $k_3 > 0$, $k_4 > 0$. Denote the concentrations of C_1, C_2, and C_3, respectively, by x_1, x_2, and x_3. Only nonnegative concentrations are physically possible. Such a reaction network gives rise to a dynamical system, which describes how the state of the network changes over time.

Suppose the dynamics of both reactions are dictated by the mass action principle. This leads to the model

$$\dot{x}_1 = -k_1 x_1^\alpha + k_2 x_2^\alpha$$
$$\dot{x}_2 = k_1 x_1^\alpha - k_2 x_2^\alpha - k_3 x_2^\alpha + k_4 x_3^\alpha$$
$$\dot{x}_3 = k_3 x_2^\alpha - k_4 x_3^\alpha$$

where $\alpha \geq 1$ is an integer. It is of primary importance in biochemical reaction network to find out the existence of multiple fixed points and once that is established, it is necessary to exploit their asymptotic behaviors given the nonlinearity of the equations and the complexity of the networks.

1.2 What Is in This Monograph

The monograph consists of five parts and eight chapters. The main content is essentially from our work [77–84, 88]. The monograph concentrates on a class of continuous-time coupled dynamic systems. The objective is to ensure the asymptotic coincidence of all states of the subsystems. These systems are often encountered in physics, biology, and engineering applications such as synchronization, swarming, and multi-vehicle cooperation. By formulating in a mathematical way, we explore two issues from the structure point of view: stability analysis and distributed control. Rather than focusing on a single application area, we consider a general formalism for such problems and perform a systematic inquiry. We investigate coupled cell systems with both static (time-invariant) coupling structure and dynamic (time-varying) coupling structure.

A brief description of the contents of each chapter follows.

Chapter 2 collects relevant common notions in graph theory and defines several new terminologies in order to better suit our development. A very important result on connectivity properties of digraphs is formulated. The main purpose of this chapter is to provide a mathematical foundation based on the theory of graphs.

Chapter 3 concentrates on the deeper connections between nonnegative matrices and directed graphs. Structural properties of nonnegative matrices are established in terms of beautiful graph theoretic interpretations.

Chapter 4 focuses on the deeper connections between a class of spacial matrices called *generator matrices* and graph theory. Algebraic properties and stability properties of generator matrices are then exploited.

Chapter 5 investigates linear coupled cell systems in continuous time. Both fixed topology and dynamic topology for linear coupled cell systems are treated. For the dynamic topology case, symmetric coupling structure and asymmetric coupling structure are explored separately using different methods. Necessary and sufficient conditions for achieving agreement are derived for each case. Stability properties are also discussed.

Chapter 6 studies nonlinear coupled cell systems which are the generalization and extension of linear coupled cell systems developed in Chapter 5. Some reasonable cooperative assumptions on the vector fields of individual systems are imposed and then necessary and sufficient conditions are obtained for both static and dynamic interaction cases. In addition, synchronization of coupled oscillator, biochemical reaction network, and water tank network are discussed within the same framework of nonlinear coupled cell systems.

Chapter 7 develops distributed control for collective point-mass robots. Formation evolution under cyclic pursuit, formation control under state-independent information flow graph, and rendezvous problem under state-dependent information flow graph are studied, respectively.

Chapter 8 formulates the formation stabilization problem for coupled kinematic unicycles. Under the assumption that the information flow graph is fixed, necessary and/or sufficient conditions for the feasibility of rendezvous, row straightening, and pattern formations are presented. Then explicit distributed control is constructed to solve these problems.

Part II

Directed Graphs and Matrices

2

Connectivity in Graph Theory

In this chapter we begin by surveying some basic notions from graph theory [11,44] and then develop a very important result about connectivity properties of digraphs. The main purpose of this chapter is to provide a mathematical foundation, based on the theory of graphs. We shall strive for rigor in presentation and shall not discuss the applicability of the concepts in the real world. This is postponed for the later chapters where we apply the results developed here to various aspects of system analysis.

2.1 Digraphs, Neighbors, Degrees

A *directed graph* (or just *digraph*) $\mathcal{G}$ consists of a non-empty finite set $\mathcal{V}$ of elements called *nodes* and a finite set $\mathcal{E}$ of ordered pairs of nodes called *arcs* (see Fig. 2.1). We call $\mathcal{V}$ the *node set* and $\mathcal{E}$ the *arc set* of $\mathcal{G}$. We will often write $\mathcal{G} = (\mathcal{V}, \mathcal{E})$, which means that $\mathcal{V}$ and $\mathcal{E}$ are the node set and arc set of $\mathcal{G}$, respectively.

Fig. 2.1. Digraphs.

For an arc (u, v) the first node u is its *tail* and the second node v is its *head*. We also say that the arc (u, v) *leaves* u and *enters* v. The head and tail of an arc are its *end-nodes*. A *loop* is an arc whose end-nodes are the same node. An arc is *multiple* if there is another arc with the same end-nodes. A digraph is *simple* if it has no multiple arcs or loops.

For example, consider the digraphs represented in Fig. 2.1. Here, digraph (a) is simple; digraph (b) has multiple arcs, namely, (v_3, v_1); and digraph (c) has a loop, namely, (v_2, v_2).

In what follows, unless otherwise specified, a digraph $\mathcal{G} = (\mathcal{V}, \mathcal{E})$ is always assumed to be simple.

The local structure of a digraph is described by the neighborhoods and the degrees of its nodes. For a digraph $\mathcal{G} = (\mathcal{V}, \mathcal{E})$ and a node v in $\mathcal{V}$, we use the following notation:

$$\mathcal{N}_v^+ = \{u \in \mathcal{V} - \{v\} : (v, u) \in \mathcal{E}\}, \quad \mathcal{N}_v^- = \{u \in \mathcal{V} - \{v\} : (u, v) \in \mathcal{E}\}.$$

The sets $\mathcal{N}_v^+$ and $\mathcal{N}_v^-$ are called the *out-neighborhood* and *in-neighborhood* of v, respectively. We call the nodes in $\mathcal{N}_v^+$ and $\mathcal{N}_v^-$ the *out-neighbors* and *in-neighbors* of v. The *out-degree*, d_v^+, of a node v is the cardinality of $\mathcal{N}_v^+$. Correspondingly, the *in-degree*, d_v^-, of a node v is the cardinality of $\mathcal{N}_v^-$. In symbols, $d_v^+ = |\mathcal{N}_v^+|$ and $d_v^- = |\mathcal{N}_v^-|$.

As an illustration, consider digraph (a) in Fig. 2.1, in which we have, for the node v_1,

$$\mathcal{N}_{v_1}^+ = \{v_2, v_3\}, \quad \mathcal{N}_{v_1}^- = \{v_4\}, \quad \text{and} \quad d_{v_1}^+ = 2, \quad d_{v_1}^- = 1.$$

2.2 Walks, Paths, Cycles

A *walk* in a digraph $\mathcal{G}$ is an alternating sequence

$$\mathcal{W} : v_1 e_1 v_2 e_2 \cdots e_{k-1} v_k$$

of nodes v_i and arcs e_i such that $e_i = (v_i, v_{i+1})$ for every $i = 1, 2, \ldots, k-1$. We say that $\mathcal{W}$ is a walk *from* v_1 to v_k. The *length* of a walk is the number of its arcs. Hence the walk $\mathcal{W}$ above has length $k - 1$. A *semiwalk* in a digraph $\mathcal{G}$ is an alternating sequence

$$v_1 e_1 v_2 e_2 \cdots e_{k-1} v_k$$

of nodes and arcs such that $e_i = (v_i, v_{i+1})$ or $e_i = (v_{i+1}, v_i)$ for every $i = 1, 2, \ldots, k-1$.

If the nodes of a walk $\mathcal{W}$ are distinct, $\mathcal{W}$ is a *path*. If the nodes $v_1, \ldots, v_{k-1}$ are distinct and $v_1 = v_k$, $\mathcal{W}$ is a *cycle*. Since paths and cycles are special cases of walks, the *length* of a path and a cycle is already defined. Cycles of length 1 are *loops*. A digraph without cycles is said to be *acyclic*.

These concepts are now illustrated. For the digraph in Fig. 2.2,

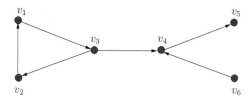

Fig. 2.2. Walk, semiwalk, path, and cycle.

$$v_1(v_1, v_3)v_3(v_3, v_4)v_4(v_4, v_5)v_5$$

is not only a walk but also a path from v_1 to v_5, and

$$v_1(v_1, v_3)v_3(v_3, v_2)v_2(v_2, v_1)v_1$$

is not only a walk from v_1 to v_1 but also a cycle. However,

$$v_1(v_1, v_3)v_3(v_3, v_2)v_2(v_2, v_1)v_1(v_1, v_3)v_3$$

is just a walk from v_1 to v_3 which is neither a path nor a cycle. In addition,

$$v_1(v_1, v_3)v_3(v_3, v_4)v_4(v_6, v_4)v_6$$

is not a walk but a semiwalk from v_1 to v_6. Nevertheless, all the walks are semiwalks.

When a digraph $\mathcal{G}$ is not acyclic, the *period d* of $\mathcal{G}$ is defined as the greatest common divisor of all the lengths of cycles in $\mathcal{G}$. We call the digraph *d-periodic* if $d > 1$ and *aperiodic* if $d = 1$. For each node v_i in $\mathcal{G}$, let S_i be the set of all the lengths, m_i^k, of walks from v_i to v_i and define

$$d_i = \operatorname*{g.c.d.}_{m_i^k \in S_i} \{m_i^k\},$$

the greatest common divisor of all the lengths, the *period* of the node v_i. We call
the node v_i d_i-*periodic* if $d_i > 1$ and *aperiodic* if $d_i = 1$. When we say that a node
v_i is d_i-periodic or aperiodic, there has to be a cycle through v_i. As an example,

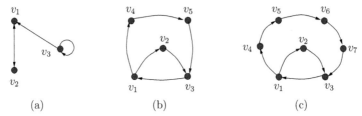

Fig. 2.3. Aperiodic and d-periodic digraphs.

three digraphs are given in Fig. 2.3. Clearly, the digraph (a) is aperiodic since
it has a loop. For the digraph (b), there are two cycles: one of them is of length
3 and the other is of length 4, so it is also aperiodic. However, for the digraph
(c), there are still two cycles but the lengths are 3 and 6, respectively, so it is
3-periodic. Now we look at the node v_1 in the digraph (a). The lengths of walks
from v_1 to v_1 are 2, 4, 6, ..., meaning that the walk starts at v_1 and ends at
v_1 but it could repeatedly traverse v_2 and v_1 for any positive integer times, so
the node v_1 has a period 2, which is different from the period of the digraph.
However, if the digraph is strongly connected, then the period of every node is
the same as the period of the digraph, which we will show later on. For example,
in the digraph (c), we choose any node, say v_4. The lengths of walks from v_4
to v_4 are 6, 9, 12, 15, ..., which are actually nonnegative combinations of the
lengths of two cycles, 3 and 6. Hence, the node v_4 is of period 3, equaling the
period of the digraph.

2.3 Connectedness

One of the most important graph theoretic concepts is that of *connectedness*. We
now introduce some of the ideas concerned with this aspect of digraph structure.

For a digraph $\mathcal{G}$, if there is a walk from one node u to another node v, then
v is said to be *reachable* from u, written $u \to v$. If not, then v is said to be not
reachable from u, written $u \nrightarrow v$. In particular, a node v is reachable from itself
by recalling that the sequence v is a trivial walk of length 0.

A node v which is reachable from every node of the digraph $\mathcal{G}$ is called a *globally reachable node* of the digraph. A node v from which every node of the digraph $\mathcal{G}$ is reachable is called a *center node* of the digraph.

A digraph $\mathcal{G}$ is *fully connected* if for every two nodes u and v there are an arc from u to v and an arc from v to u; $\mathcal{G}$ is *strongly connected* if every two nodes u and v are mutually reachable; $\mathcal{G}$ is *unilaterally connected* if for every two nodes u and v at least one is reachable from the other; $\mathcal{G}$ is *quasi-strongly connected (QSC)* if for every two nodes u and v there is a node w from which u and v are reachable; $\mathcal{G}$ is *weakly connected* if every two nodes u and v are joined by a semiwalk (disregarding the orientation of each arc). A digraph $\mathcal{G}$ is *disconnected* if it is not even weakly connected. It is easy to see that $\mathcal{G}$ is strongly connected if and only if every node of $\mathcal{G}$ is a globally reachable node, or equivalently every node of $\mathcal{G}$ is a center node. Clearly, a digraph consisting of only one node is always strongly connected since the node is reachable from itself.

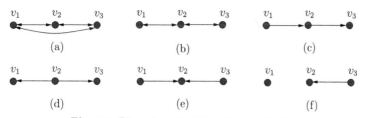

Fig. 2.4. Digraphs with different connectivity.

Fig. 2.4 shows: (a) a fully connected digraph, (b) a strongly connected digraph, (c) a unilaterally connected digraph, (d) a quasi-strongly connected digraph, (e) a weakly connected digraph, (f) a disconnected digraph.

Clearly, every fully connected digraph is strongly connected, every strongly connected digraph is unilaterally connected, every unilaterally connected digraph is quasi-strongly connected, and every quasi-strongly connected digraph is weakly connected, but the converses of these statements are not true in general. Hence these kinds of connectedness for digraphs are overlapping.

2.4 Operations on Digraphs

We first study the operation of taking the "converse" of any digraph. We shall
see that this operation, which involves reversing the direction of every arc of a
given digraph, sets the stage for a powerful principle called "directional duality".
This principle will enable us to establish certain theorems without effort once we
have proved other corresponding theorems. Note that the digraphs of Fig. 2.5
(a) and (b) are related to each other in a particular way: either one can be
obtained from the other simply by reversing the directions of all arcs. Given a
digraph $\mathcal{G}$, its *opposite digraph* $\mathcal{G}^*$ is the digraph with the same node set formed
by exchanging the orientations of all arcs in $\mathcal{G}$.

(a)   (b)

Fig. 2.5. Digraph and its opposite digraph.

Next we study the operation of taking the "union" of any two or more digraphs
which have the same node set. If $\mathcal{G} = (\mathcal{V}, \mathcal{E})$ and $\mathcal{G}' = (\mathcal{V}, \mathcal{E}')$ are digraphs with
the same node set $\mathcal{V}$, then their *union* $\mathcal{G} \cup \mathcal{G}'$ is the digraph with arc set $\mathcal{E} \cup \mathcal{E}'$.
That is,

$$\mathcal{G} \cup \mathcal{G}' = (\mathcal{V}, \mathcal{E} \cup \mathcal{E}') .$$

Fig. 2.6 provides an example of union operation for two digraphs, $\mathcal{G}$ and $\mathcal{G}'$, with
the same node set $\{v_1, v_2, v_3\}$.

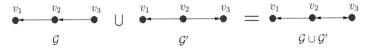

Fig. 2.6. Digraphs and their union.

Finally, it is sometimes appropriate to examine just part of a digraph. This can
be done in the following way. For a digraph $\mathcal{G} = (\mathcal{V}, \mathcal{E})$, if $\mathcal{U}$ is a nonempty subset
of $\mathcal{V}$, then the digraph $(\mathcal{U}, \mathcal{E} \cap (\mathcal{U} \times \mathcal{U}))$ is termed the *induced subdigraph* by $\mathcal{U}$.
A *strong component* of a digraph $\mathcal{G} = (\mathcal{V}, \mathcal{E})$ is a maximal induced subdigraph

of $\mathcal{G}$ which is strongly connected (maximal induced subdigraph is not unique in general). If $\mathcal{G}_1 = (\mathcal{V}_1, \mathcal{E}_1), \ldots, \mathcal{G}_k = (\mathcal{V}_k, \mathcal{E}_k)$ are the strong components of $\mathcal{G} = (\mathcal{V}, \mathcal{E})$, then clearly $\mathcal{V}_1 \cup \cdots \cup \mathcal{V}_k = \mathcal{V}$ (recall that a digraph with only one node is strongly connected). Moreover, we must have $\mathcal{V}_i \cap \mathcal{V}_j = \phi$ for every $i \neq j$ as otherwise all the nodes in $\mathcal{V}_i \cup \mathcal{V}_j$ are reachable from each other, implying that the nodes of $\mathcal{V}_i \cup \mathcal{V}_j$ belong to the same strong component of $\mathcal{G}$. In other words, every node belongs to exactly one strong component of $\mathcal{G}$.

On the other hand, for a digraph $\mathcal{G} = (\mathcal{V}, \mathcal{E})$, a nonempty node set $\mathcal{U} \subseteq \mathcal{V}$ is *closed* if the node v is not reachable from u for all $u \in \mathcal{U}$ and $v \in \mathcal{V} - \mathcal{U}$. In other words, there is no arc leaving from the node set $\mathcal{U}$. In particular, $\mathcal{U} = \mathcal{V}$ is closed. A strong component $\mathcal{G}_1 = (\mathcal{V}_1, \mathcal{E}_1)$ of a digraph $\mathcal{G}$ is *closed* if $\mathcal{V}_1$ is closed in $\mathcal{G}$. In fact, an induced subdigraph having a minimal closed node subset in $\mathcal{G}$ is a strong component of $\mathcal{G}$.

Fig. 2.7 provides examples of induced subdigraphs, $\mathcal{G}_1$, $\mathcal{G}_2$, and $\mathcal{G}_3$, of the first digraph $\mathcal{G}$, where $\mathcal{G}_1$ is not a strong component, $\mathcal{G}_2$ is a strong component but it is not closed, and $\mathcal{G}_3$ is a closed strong component.

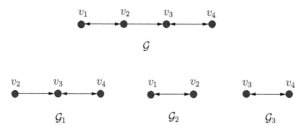

Fig. 2.7. Induced subdigraph, strong component, and closed strong component.

2.5 Dynamic Graphs

We introduce the notion of dynamic graph, which is a digraph whose connectivity changes over time. Consider a set of n nodes, $\mathcal{V}$, and a set of all possible arc sets with n nodes, $\{\mathcal{E}_p, \ p \in \mathcal{Q}\}$, where $\mathcal{Q}$ is the set of indices. Denote the set of all possible digraphs with n nodes by $2^{\mathcal{G}} = \{(\mathcal{V}, \mathcal{E}_p) : p \in \mathcal{Q}\}$. A *dynamic graph* $\mathcal{G}(t) : \mathbb{R} \to 2^{\mathcal{G}}$ is a graph defined on the time axis.

Given a dynamic graph $\mathcal{G}(t) = (\mathcal{V}, \mathcal{E}(t))$, we denote by $\mathcal{G}([t_1, t_2])$ the union digraph whose arcs are obtained from the union of the arcs in $\mathcal{G}(t)$ over the time interval $[t_1, t_2]$, that is,

$$\mathcal{G}([t_1, t_2]) = \left(\mathcal{V}, \bigcup_{t \in [t_1, t_2]} \mathcal{E}(t) \right).$$

Also, it is convenient to introduce a very important concept here, which is a property of connectedness for dynamic graphs.

Definition 2.1. *A dynamic graph $\mathcal{G}(t)$ is* uniformly quasi-strongly connected (UQSC) (uniformly strongly connected) *if there exists $T > 0$ such that for all t, the union digraph $\mathcal{G}([t, t + T])$ is quasi-strongly connected (strongly connected).*

The uniformity in the definition above means the time length T should be independent of the starting time t of any interval.

For example, consider a dynamic graph $\mathcal{G}(t)$ defined in Fig. 2.8, which pe-

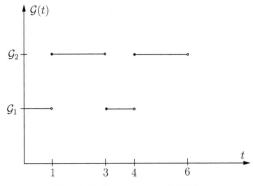

Fig. 2.8. Dynamic graph $\mathcal{G}(t)$.

riodically switches between $\mathcal{G}_1$ and $\mathcal{G}_2$ shown in Fig. 2.9. It can be easily verified that there is a $T = 3$ (the period) such that for all t, the union digraph $\mathcal{G}([t, t + T]) = \mathcal{G}_1 \cup \mathcal{G}_2$ is quasi-strongly connected. Hence the dynamic graph $\mathcal{G}(t)$ is uniformly quasi-strongly connected.

Instead, if $\mathcal{G}(t)$ is defined like the following:

Fig. 2.9. Digraphs $\mathcal{G}_1$ and $\mathcal{G}_2$.

$$\mathcal{G}(t) = \begin{cases} \mathcal{G}_1, & \text{for } t \in [0,1) \cup [2^k, 2^{k+1}), \ k = 1, 3, 5, \ldots \\ \mathcal{G}_2, & \text{for } t \in [2^k, 2^{k+1}), \ k = 0, 2, 4, \ldots \end{cases}$$

it is not uniformly quasi-strongly connected as we are unable to find such T.

2.6 Undirected Graphs

For completeness, we review some concepts for undirected graphs. Undirected graphs form in a sense a special class of directed graphs (symmetric digraphs) and hence problems that can be formulated for both directed and undirected graphs are often easier for the latter.

An *undirected graph* $\mathcal{G} = (\mathcal{V}, \mathcal{E})$ consists of a non-empty finite set $\mathcal{V}$ of elements called *nodes* and a finite set $\mathcal{E}$ of unordered pairs of nodes called *edges*.

We can simply treat an undirected graph $\mathcal{G}$ as a *bidirectional digraph* by replacing each edge (u, v) of $\mathcal{G}$ with the pair of arcs (u, v) and (v, u). Thus, they are completely the same in the sense of connectedness. An example is given in Fig. 2.10.

Undirected graph   Bidirectional digraph

Fig. 2.10. Undirected graph and bidirectional digraph.

Furthermore, it is worth pointing out that for bidirectional digraphs, the four kinds of connectedness we introduced in the previous section, namely, strongly connected, unilaterally connected, quasi-strongly connected, and weakly connected, are equivalent, and they are all referred to as *connected* in the context of undirected graphs.

2.7 Equivalence on Connectivity

In this section we present and prove a fundamental result on connectedness of digraphs, which will become extremely important in the later chapters.

Theorem 2.1. *For a digraph* $\mathcal{G} = (\mathcal{V}, \mathcal{E})$*, the following are equivalent:*

(a) *The digraph* $\mathcal{G}$ *is quasi-strongly connected;*

(b) *The digraph* $\mathcal{G}$ *has a center node;*

(c) *The opposite digraph* $\mathcal{G}^*$ *has a globally reachable node;*

(d) *The opposite digraph* $\mathcal{G}^*$ *has only one closed strong component.*

In the above theorem, the conditions (a), (b), and (c) are equivalent because the same property is stated using different terminologies. The condition (d) provides a new useful characterization of this property. Due to lack of an appropriate term in graph theory describing this property, we first introduced the notion of a *globally reachable node* and then proved the equivalence of the conditions (c) and (d), which is also proved independently in [96] with logically contrapositive form. Later on, we found the notion of *quasi-strongly connected* in [17] and became aware that the equivalent conditions (a) and (b) presented in [17] are just directional dual properties of (c) and (d). The proof of the equivalence of (a) and (b) can be found in [17], page 133, and so is omitted. In order to prove the remaining, the next preliminary result is needed, which shows the existence of a closed strong component for any digraph. The proof of the lemma also provides an algorithm to find a closed strong component.

Lemma 2.1. *A digraph* $\mathcal{G} = (\mathcal{V}, \mathcal{E})$ *has at least one closed strong component. Furthermore, if a nonempty set* $\mathcal{U} \subset \mathcal{V}$ *is closed in* $\mathcal{G}$*, then* $\mathcal{G}$ *has a closed strong component* $\mathcal{G}_c = (\mathcal{V}_c, \mathcal{E}_c)$ *satisfying* $\mathcal{V}_c \subseteq \mathcal{U}$*.*

Proof: We prove the first assertion by means of a constructive algorithm.

Select any node, say v_1, in $\mathcal{V}$. Let $\mathcal{V}_1$ be the set of nodes from which v_1 is reachable and let $\mathcal{V}_1'$ be the set of nodes which are reachable from v_1. Recall that every node is reachable from itself. So both $\mathcal{V}_1$ and $\mathcal{V}_1'$ contain element v_1.

Check whether $\mathcal{V}_1' \subseteq \mathcal{V}_1$.

If so, then the induced subdigraph $\mathcal{G}_1$ by $\mathcal{V}_1'$ is a closed strong component of $\mathcal{G}$. To see this, firstly, notice that every two nodes $u, v \in \mathcal{V}_1' \subseteq \mathcal{V}_1$ are mutually reachable since $u \to v_1 \to v$ and $v \to v_1 \to u$. So the induced subdigraph $\mathcal{G}_1$ by $\mathcal{V}_1'$ is strongly connected. On the other hand, for all $v \in \mathcal{V}_1'$ and $u \in \mathcal{V} - \mathcal{V}_1'$, $v \not\to u$ since otherwise $v_1 \to v \to u$ and $u \in \mathcal{V}_1'$. Hence, $\mathcal{V}_1'$ is closed, and the induced subdigraph by $\mathcal{V}_1' + \{u\}$ is not strongly connected since u is not reachable from any other node in $\mathcal{V}_1'$. Therefore, $\mathcal{G}_1$ is a maximal induced subdigraph which is strongly connected. In conclusion, $\mathcal{G}_1$ is a closed strong component.

If instead the condition above is false, select any node, say v_2, in $\mathcal{V} - \mathcal{V}_1$. Let $\mathcal{V}_2$ be the set of nodes from which v_2 is reachable and let $\mathcal{V}_2'$ be the set of nodes which are reachable from v_2. Thus $\mathcal{V}_2'$ must be a subset of $\mathcal{V} - \mathcal{V}_1$ since otherwise v_1 is reachable from v_2. Check whether $\mathcal{V}_2' \subseteq \mathcal{V}_2$. If so, the induced subdigraph by $\mathcal{V}_2'$ is a closed strong component of $\mathcal{G}$ by the same argument as above. If it is not, repeat this procedure again until this condition holds. The digraph $\mathcal{G}$ has a finite number of nodes and $\mathcal{V}_k'$ is getting smaller each step by noting that

$$\mathcal{V}_k' \subseteq \mathcal{V} - \mathcal{V}_1 - \cdots - \mathcal{V}_{k-1}.$$

So eventually the condition must hold. Indeed, $\mathcal{V}_m'$ has only one element v_m at some step m if the condition is not satisfied before step m. Thus $\mathcal{V}_m' = \{v_m\} \subseteq \mathcal{V}_m$ by recalling that v_m is also an element of $\mathcal{V}_m$. Therefore the closed strong component of $\mathcal{G}$ has been constructed.

If a nonempty node set $\mathcal{U} \subset \mathcal{V}$ is closed in $\mathcal{G}$, we let the induced subdigraph by $\mathcal{U}$ be $\mathcal{G}_u = (\mathcal{U}, \mathcal{E}_u)$. By the first assertion, we know that the digraph $\mathcal{G}_u$ has at least one closed strong component, say $\mathcal{G}_c = (\mathcal{V}_c, \mathcal{E}_c)$. Obviously $\mathcal{V}_c \subseteq \mathcal{U}$. It remains to show that $\mathcal{G}_c$ is also a closed strong component of $\mathcal{G}$. Clearly, $\mathcal{G}_c$ is also an induced subdigraph of $\mathcal{G}$. Moreover, $\mathcal{V}_c$ is closed in $\mathcal{G}$. Therefore, $\mathcal{G}_c$ is the maximal induced subdigraph which is strongly connected. Combining with the fact that $\mathcal{V}_c$ is closed in $\mathcal{G}$, it follows that $\mathcal{G}_c$ is a closed strong component of the digraph $\mathcal{G}$. ∎

Proof of Theorem 2.1: (b) $\Longleftrightarrow$ (c) By the definition of opposite digraph, immediately we know that a center node of the digraph $\mathcal{G}$ is a globally reachable node of the opposite digraph $\mathcal{G}^*$.

(c) $\Longrightarrow$ (d) If the opposite digraph $\mathcal{G}^*$ has a globally reachable node, let $\mathcal{V}_1$ be the subset of $\mathcal{V}$ consisting all the globally reachable nodes and let $\mathcal{G}_1$ be the induced subdigraph by $\mathcal{V}_1$, then we claim that $\mathcal{G}_1$ is the only closed strong component in $\mathcal{G}^*$. The set $\mathcal{V}_1$ may equal $\mathcal{V}$ or be a strict subset of $\mathcal{V}$. In the first case, $\mathcal{V}_1 = \mathcal{V}$, clearly $\mathcal{G}_1 = \mathcal{G}^*$ is strongly connected and so it is the unique closed strong component of $\mathcal{G}^*$. In the second case, $\mathcal{V}_1 \subset \mathcal{V}$, we have that v is not reachable from u for all $u \in \mathcal{V}_1$ and $v \in \mathcal{V} - \mathcal{V}_1$, implying $\mathcal{V}_1$ is closed. (To see this point, suppose by contradiction that there are $u \in \mathcal{V}_1$ and $v \in \mathcal{V} - \mathcal{V}_1$ such that v is reachable from u. Notice that u is a globally reachable node. So v is also globally reachable, which contradicts that $v \notin \mathcal{V}_1$ is not a globally reachable node.)

For any two distinct nodes $u, v \in \mathcal{V}_1$, there is a walk from u to v in $\mathcal{G}^*$ since both nodes are globally reachable. Furthermore, since $\mathcal{V}_1$ is closed, this walk cannot go through any node not in $\mathcal{V}_1$ and must be in the induced subdigraph $\mathcal{G}_1$. That means $\mathcal{G}_1$ is strongly connected.

Moreover, since $\mathcal{V}_1$ is closed and no node in $\mathcal{V} - \mathcal{V}_1$ is reachable from any node in $\mathcal{V}_1$, no node can be added to make the induced subdigraph strongly connected. This implies $\mathcal{G}_1$ is the maximal induced subdigraph which is strongly connected. Hence it is a closed strong component of $\mathcal{G}^*$.

Finally we show it is the unique one in $\mathcal{G}^*$. Suppose by contradiction that there is another closed strong component in $\mathcal{G}^*$, say $\mathcal{G}_2 = (\mathcal{V}_2, \mathcal{E}_2)$. Recall that $\mathcal{V}_1 \cap \mathcal{V}_2 = \phi$. Since $\mathcal{V}_2$ is closed by assumption, for any node $v \in \mathcal{V}_1$ and any node $u \in \mathcal{V}_2$, v is not reachable from u, which contradicts the fact that v is a globally reachable node.

(c) $\Longleftarrow$ (d) If the opposite digraph $\mathcal{G}^*$ has only one closed strong component, say $\mathcal{G}_1 = (\mathcal{V}_1, \mathcal{E}_1)$, we claim that every node in $\mathcal{V}_1$ is globally reachable. Suppose by contradiction that there is a node $v \in \mathcal{V}_1$ which is not globally reachable. Let $\mathcal{V}_2$ be the set of nodes from which v is reachable and let $\mathcal{V}_3$ be the set of nodes from which v is not reachable. Then for any node $u \in \mathcal{V}_2$ and any node $w \in \mathcal{V}_3$, $w \nrightarrow u$ since otherwise $w \rightarrow u \rightarrow v$. Notice that $\mathcal{V}_2 \cup \mathcal{V}_3 = \mathcal{V}$. So it follows that $\mathcal{V}_3$ is closed. Let $\mathcal{G}_3 = (\mathcal{V}_3, \mathcal{E}_3)$ be the induced subdigraph by $\mathcal{V}_3$. Then $\mathcal{G}_3$ has a closed strong component by Lemma 2.1, which is also a closed strong component of $\mathcal{G}^*$ since $\mathcal{V}_3$ is closed. Furthermore, this closed strong component is not the same as $\mathcal{G}_1$ since it does not have the node v while $\mathcal{G}_1$ has the node v. Therefore, $\mathcal{G}^*$ has two closed strong components, a contradiction. ∎

As an illustration, Fig. 2.11 shows a digraph $\mathcal{G}$ with six nodes and its opposite digraph $\mathcal{G}^*$. It can be checked that $\mathcal{G}$ is quasi-strongly connected and it has two center nodes, namely, 3 and 4. The opposite digraph $\mathcal{G}^*$ has two global reachable nodes, 3 and 4. Moreover, $\mathcal{G}^*$ has only one strong component, namely, the induced subdigraph by the node set $\{3, 4\}$.

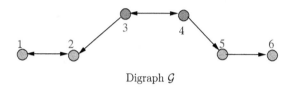

Digraph $\mathcal{G}$

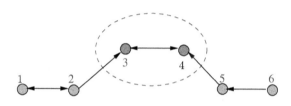

The opposite digraph $\mathcal{G}^*$

Fig. 2.11. Equivalence on connectivity.

3

Nonnegative Matrices and Graphs

In this chapter we explore the theory of nonnegative matrices with emphasis on the deeper connections between nonnegative matrices and directed graphs. An excellent reference on nonnegative matrices is [18]. We develop several new results on this topic, which can be treated as a complement to the results in [18] for independent interest.

We shall deal with square *nonnegative matrices* $E = (e_{ij})$, $i, j = 1, \ldots, n$; i.e., $e_{ij} \geq 0$ for all i, j, in which case we write $E \succeq 0$. If, in fact, $e_{ij} > 0$ for all i, j, we shall put $E \succ 0$ and call it *positive*.

This definition and notation apply to row vectors x^T and column vectors x. We shall use the notation $E_1 \succeq E_2$ to mean $E_1 - E_2 \succeq 0$, where E_1, E_2, and 0 are square nonnegative matrices of compatible dimensions.

The approach in the main body of this chapter is combinatorial, using the element-wise structure in which the zero-nonzero pattern plays an important role. The zero-nonzero pattern of a nonnegative matrix is completely determined by an associated digraph. So in the chapter we shall study the structural properties of nonnegative matrices which have beautiful graph theoretic interpretations.

Finally, we shall use the notation $E^k = (e_{ij}^{(k)})$ for the kth power of E and use the notation $\rho(E)$ for the spectral radius of E, which is the maximum modulus of the eigenvalues of E.

3.1 Adjacency Matrices and Digraphs

We start by defining the associated digraph of a nonnegative matrix.

For an $n \times n$ nonnegative matrix E, the *associated digraph* $\mathcal{G}(E)$ consists of n nodes $v_1, \ldots, v_n$ where an arc leads from v_i to v_j if and only if $e_{ij} \neq 0$.

It is clear that if any other nonnegative matrix $\bar{E}$ has the same dimensions as E, and has positive entries and zero entries in the same positions as E, then they have the same associated digraph. If two nonnegative matrices E and $\bar{E}$ have the same associated digraph, we will say that they are of the same structure, written as $E \sim \bar{E}$.

From the viewpoint of graph theory, there is a typical nonnegative matrix associated with a digraph, which is called its adjacency matrix. An *adjacency matrix* $\bar{E} = (\bar{e}_{ij}) \in \mathbb{R}^{n \times n}$ of a digraph $\mathcal{G}$, with n nodes, is the matrix in which $\bar{e}_{ij} = 1$ if there is an arc leading from the node v_i to the node v_j in $\mathcal{G}$ and $\bar{e}_{ij} = 0$ otherwise.

Thus, an adjacency matrix $\bar{E}$ corresponding to a given nonnegative matrix E replaces all the positive entries of E by ones and of course $\bar{E} \sim E$.

We can make a number of observations about the adjacency matrix of a digraph:

(a) $\bar{E}$ is not necessarily symmetric.

(b) The sum of the entries in row i of $\bar{E}$ is equal to the out-degree of v_i.

(c) The sum of the entries in column j of $\bar{E}$ is equal to the in-degree of v_j.

As an example, let

$$E = \begin{pmatrix} 0 & 5 & 0 & 0 \\ 2 & 0 & 0 & 6 \\ 0 & 3 & 0 & 0 \\ 1.5 & 0 & 0.5 & 0 \end{pmatrix}. \tag{3.1}$$

Then the associated digraph and its adjacency matrix are shown in Fig. 3.1.

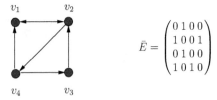

$$\bar{E} = \begin{pmatrix} 0 & 1 & 0 & 0 \\ 1 & 0 & 0 & 1 \\ 0 & 1 & 0 & 0 \\ 1 & 0 & 1 & 0 \end{pmatrix}$$

Fig. 3.1. Associated digraph and adjacency matrix.

Finally, we bring in the Gershgorin disk theorem that will be quite useful to explore the eigenvalues distribution of nonnegative matrices.

For a square matrix $E = (e_{ij})$, around every entry e_{ii} on the diagonal of the matrix, draw a closed disc of radius $\sum_{j \neq i} |e_{ij}|$. Such discs are called Gershgorin discs.

Theorem 3.1 ([22], Gershgorin, 1931). *Every eigenvalue of E lies in a Gershgorin disc.*

As an example, for the matrix E in (3.1), the Gershgorin discs are drawn in Fig. 3.2 and the eigenvalues lie in the union of these discs, i.e., the largest disc.

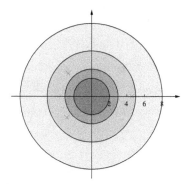

Fig. 3.2. Gershgorin discs.

3.2 Irreducible Matrices

A matrix E is *cogredient* [18] to a matrix $\tilde{E}$ if there is a permutation matrix P such that $PEP^T = \tilde{E}$. An $n \times n$ matrix E is *reducible* if $n = 1$ or, when $n > 1$, if it is cogredient to

$$\tilde{E} = \begin{pmatrix} B & 0 \\ C & D \end{pmatrix}, \tag{3.2}$$

where B and D are nonempty square matrices. Otherwise, E is *irreducible*.

By a permutation operation for a nonnegative matrix, the associated digraph only changes by a renumbering of the nodes. For example, given a nonnegative matrix E and a permutation matrix P as follows:

$$E = \begin{pmatrix} 0\,2\,1 \\ 2\,0\,0 \\ 0\,3\,0 \end{pmatrix}, \quad P = \begin{pmatrix} 0\,1\,0 \\ 0\,0\,1 \\ 1\,0\,0 \end{pmatrix},$$

then

$$\tilde{E} = PEP^T = \begin{pmatrix} 0\,0\,2 \\ 3\,0\,0 \\ 2\,1\,0 \end{pmatrix}.$$

The associated digraphs $\mathcal{G}(E)$ and $\mathcal{G}(\tilde{E})$ are given in Fig. 3.3 and we can see that they can be mutually obtained by renumbering the nodes.

$$\mathcal{G}(E) \qquad\qquad \mathcal{G}(\tilde{E})$$

Fig. 3.3. Renumbering the nodes.

Theorem 3.2 ([18], page 30). *An $n \times n$ nonnegative matrix E is irreducible if and only if $\mathcal{G}(E)$ is strongly connected.*

Let

$$E_1 = \begin{pmatrix} 0\,2\,0\,0 \\ 0\,0\,3\,0 \\ 0\,0\,0\,4 \\ 5\,0\,0\,0 \end{pmatrix} \quad \text{and} \quad E_2 = \begin{pmatrix} 0\,2\,0\,0 \\ 1\,0\,3\,0 \\ 0\,0\,0\,4 \\ 0\,0\,5\,0 \end{pmatrix}.$$

Then the associated digraphs, $\mathcal{G}(E_1)$ and $\mathcal{G}(E_2)$, are given in Fig. 3.4. It can be easily seen that $\mathcal{G}(E_1)$ is strongly connected so E_1 is irreducible and $\mathcal{G}(E_2)$ is not strongly connected so E_2 is reducible. Indeed, we can choose a permutation

$$P = \begin{pmatrix} 0\,0\,1\,0 \\ 0\,0\,0\,1 \\ 1\,0\,0\,0 \\ 0\,1\,0\,0 \end{pmatrix} \quad \text{such that} \quad PE_2P^T = \begin{pmatrix} 0\,4\,|\,0\,0 \\ 5\,0\,|\,0\,0 \\ 0\,0\,|\,0\,2 \\ 3\,0\,|\,1\,0 \end{pmatrix},$$

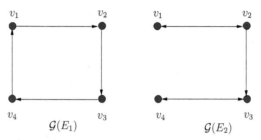

Fig. 3.4. Associated digraphs of irreducible and reducible matrices.

which is of the form in (3.2).

We end this section by invoking the following theorem which states the relationship between the structure of the k-th power of a nonnegative matrix, i.e.,

$$E^k = \left(e_{ij}^{(k)} \right),$$

and the walks in the associated digraph. This theorem is easily proved by induction on k.

Theorem 3.3 ([14], page 87). *Let E be a nonnegative matrix. Then $e_{ij}^{(k)} > 0$ if and only if $\mathcal{G}(E)$ has a walk from the node v_i to the node v_j of length k.*

3.3 Primitive Matrices

A nonnegative matrix E is said to be *primitive* [18] if there exists a positive integer k such that $E^k \succ 0$.

A primitive matrix is irreducible, but the converse is not true in general. This fact can be easily seen from the following graph characterization that we developed.

Theorem 3.4. *An $n \times n$ nonnegative matrix E is primitive if and only if $\mathcal{G}(E)$ is strongly connected and aperiodic.*

The proof requires the following lemmas.

Lemma 3.1. *Let $m_1, m_2 \geq 1$ be integers. If g.c.d$\{m_1, m_2\} = 1$, then there is an integer $\bar{k} \geq 0$ such that for any integer $k \geq \bar{k}$,*

$$k = \alpha m_1 + \beta m_2,$$

where α, β are suitable nonnegative integers.

Proof: Since

$$\text{g.c.d}\{m_1, m_2\} = 1,$$

1 is an integer combination of m_1 and m_2. Without loss of generality, say

$$1 = \alpha_1 m_1 - \beta_1 m_2, \tag{3.3}$$

where α_1, β_1 are nonnegative integers. Let $\bar{k} = \beta_1 m_2^2$. Thus $\bar{k} \geq 0$ and for all $k \geq \bar{k}$,

$$k = \beta_1 m_2^2 + i \cdot m_2 + j, \tag{3.4}$$

for some integers i, j satisfying $i \geq 0$ and $0 \leq j < m_2$. Substituting (3.3) into (3.4) leads to

$$k = \beta_1 m_2^2 + i \cdot m_2 + j \cdot (\alpha_1 m_1 - \beta_1 m_2) = (j \cdot \alpha_1) \cdot m_1 + (\beta_1(m_2 - j) + i) \cdot m_2.$$

Let

$$\alpha = j \cdot \alpha_1 \text{ and } \beta = \beta_1(m_2 - j) + i.$$

Clearly, α, β are nonnegative integers by noticing that $j < m_2$. Therefore, the conclusion follows. ∎

The next result shows the relationship between the period of a digraph and the period of each node in the digraph when it is strongly connected. This lemma is adapted from Theorem 2.2.30 in nonnegative matrix theory [18] but we present a different statement and a different proof in the context of graph theory.

Lemma 3.2. *Let d be the period of a digraph $\mathcal{G}$ and d_i be the period of node $v_i, i = 1, \ldots, n$ in $\mathcal{G}$. If $\mathcal{G}$ is strongly connected then $d = d_1 = \cdots = d_n$.*

Proof: Let $S = \{m_1, \ldots, m_p\}$ be the set of all the lengths of cycles in $\mathcal{G}$. Obviously it is a finite set by the definition of cycles and d is the greatest common divisor of S. For any node v_i, let S_i be the set of all the lengths of walks from v_i to v_i. Then d_i is the greatest common divisor of S_i. For any walk from v_i to v_i, it is either a cycle or is generated by a number of cycles (see Fig. 3.5).

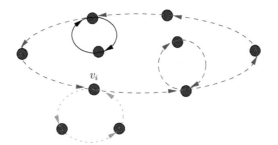

Fig. 3.5. Walk from v_i to v_i and cycles.

So the length of any walk, i.e., any element in S_i, is a linear combination of $m_j, j = 1, \ldots, p$, with nonnegative integer coefficients and therefore it divides d, the greatest common divisor of $m_j, j = 1, \ldots, p$. This further implies that d_i divides d.

On the other hand, consider any cycle in the digraph. Let the length be m_j. If it goes through v_1, then d_1 divides m_j. If not, then it has to go through some other node, say v_2. Since the digraph is strongly connected, there must be a cycle going through v_1 and v_2. Let's say the length of this cycle m_k. Thus d_1 divides m_k. Notice that these two cycles generate a walk of length $m_k + m_j$ from v_1 to v_1. So d_1 divides $m_k + m_j$ and therefore d_1 divides m_j. Hence, d_1 divides any m_j in S. That means d_1 divides d.

Hence, $d = d_1$. By the same argument, we can show $d = d_i$ for all $i = 1, \ldots, n$ if $\mathcal{G}$ is strongly connected. ■

Lemma 3.3. *Let E be an $n \times n$ nonnegative matrix. If $\mathcal{G}(E)$ is strongly connected and d-periodic, then $e_{ii}^{(k)} = 0$ for any $i = 1, \ldots, n$ and for any k that is not a multiple of d.*

Proof: Let $d_i, i = 1, \ldots, n$, be the periods of the nodes in $\mathcal{G}(E)$. Thus

$$d = d_1 = \cdots = d_n$$

by Lemma 3.2 since $\mathcal{G}(E)$ is strongly connected. Hence for any node v_i in $\mathcal{G}(E)$, the length of any walk from v_i to v_i is a multiple of d, and there is no walk from v_i to v_i with length of k that is not a multiple of d. So it follows from Theorem 3.3 that $e_{ii}^{(k)} = 0$ for any $i = 1, \ldots, n$ and any k that is not a multiple of d. ■

Proof of Theorem 3.4: ($\Longleftarrow$) If $\mathcal{G}(E)$ is strongly connected and aperiodic, then by Lemma 3.2 the period of $\mathcal{G}$ and the period of each node v_i are all equal to 1. For any node v_i, let m_i^1, m_i^2 $(m_i^1 \neq m_i^2)$ be the lengths of two walks from v_i to v_i. By Lemma 3.1 there is sufficiently large $\bar{k}_i$ such that for any $k \geq \bar{k}_i$, k can be expressed by a nonnegative integer combination of m_i^1 and m_i^2, which means there is a walk of length k from v_i to v_i. Let v_j be another node. Since there is a path from v_i to v_j, we let its length be l_{ij}. Thus for any $k \geq q_{ij} := \bar{k}_i + l_{ij}$ there is a walk of length of k from v_i to v_j. It follows from Theorem 3.3 that $e_{ij}^{(k)} > 0$ for all $k \geq q_{ij}$. Let

$$q = \max\{q_{ij} : i, j = 1, \ldots, n\}.$$

Then we have $e_{ij}^{(k)} > 0$ for all $i, j = 1, \ldots, n$ and $k \geq q$. By the definition of primitive matrix, E is primitive.

($\Longrightarrow$) To prove the contrapositive form, assume that $\mathcal{G}(E)$ is not strongly connected, or that it is strongly connected but it is not aperiodic. For the first case that $\mathcal{G}(E)$ is not strongly connected, there is a pair of nodes v_i and v_j such that v_j is not reachable from v_i. So by Theorem 3.3, $e_{ij}^{(k)} = 0$ for all $k > 0$. Hence there is no natural number k such that E^k is positive and E is not primitive.

For the second case, $\mathcal{G}(E)$ is strongly connected but it is not aperiodic, that is, it is d-periodic, where $d > 1$. So it follows from Lemma 3.3 that $e_{ii}^{(k')} = 0$ for any positive integer k' that is not a multiple of d. Hence there is no natural number k such that E^k is positive since otherwise if there is a natural number k^* such that E^{k^*} is positive, then E^k is positive for any $k \geq k^*$, which contradicts $e_{ii}^{(k')} = 0$ for any positive integer k' that is not a multiple of d. Therefore, E is not primitive. $\blacksquare$

As an example, let

$$E_1 = \begin{pmatrix} 0 & 2 & 0 \\ 1 & 0 & 0 \\ 0 & 3 & 0 \end{pmatrix}, \quad E_2 = \begin{pmatrix} 0 & 1 & 0 \\ 0 & 0 & 1 \\ 1 & 0 & 0 \end{pmatrix}, \quad \text{and} \quad E_3 = \begin{pmatrix} 0 & 1 & 0 \\ 0 & 0 & 1 \\ 1 & 1 & 0 \end{pmatrix}.$$

Then the associated digraphs, $\mathcal{G}(E_1)$, $\mathcal{G}(E_2)$, and $\mathcal{G}(E_3)$, are given in Fig. 3.6. As we can see, $\mathcal{G}(E_1)$ is not strongly connected so E_1 is reducible and so it is not primitive. The digraph $\mathcal{G}(E_2)$ is strongly connected but it is periodic of period 3 so it is irreducible but not primitive. Indeed, for any positive integer k,

v_1 v_2 v_3 v_1 v_2 v_3 v_1 v_2 v_3

$\mathcal{G}(E_1)$ $\mathcal{G}(E_2)$ $\mathcal{G}(E_3)$

Fig. 3.6. Associated digraphs of non-primitive and primitive matrices.

$$E_2^{3k-2} = \begin{pmatrix} 0\,1\,0 \\ 0\,0\,1 \\ 1\,0\,0 \end{pmatrix}, \quad E_2^{3k-1} = \begin{pmatrix} 0\,0\,1 \\ 1\,0\,0 \\ 0\,1\,0 \end{pmatrix}, \quad \text{and } E_2^{3k} = \begin{pmatrix} 1\,0\,0 \\ 0\,1\,0 \\ 0\,0\,1 \end{pmatrix}.$$

Obviously, there is no positive integer k such that the k-th power of E_2 is positive. The associated digraph $\mathcal{G}(E_3)$ is strongly connected and aperiodic ($d = $ g.c.d$\{2, 3\} = 1$) so E_3 is primitive. Indeed, when $k = 5$,

$$E_3^k = \begin{pmatrix} 1\,1\,1 \\ 1\,2\,1 \\ 1\,2\,2 \end{pmatrix}$$

is positive.

Finally we recall the well-known Perron-Frobenius theorem here, which actually makes the theory of nonnegative matrices so attractive.

Theorem 3.5 ([18], Perron-Frobenius, 1907). *Let E be a nonnegative, irreducible matrix. The following are true:*

(a) *$\rho(E)$ is a simple eigenvalue and any eigenvalue of E of the same modulus is also simple.*

(b) *The matrix E has a positive eigenvector x corresponding to $\rho(E)$.*

If, in addition, E is primitive, then all eigenvalues of E other than $\rho(E)$ have modulus less than $\rho(E)$.

3.4 Stochastic Matrices

A class of very interesting and useful nonnegative matrices are stochastic matrices. We first review some properties of stochastic matrices from [18] and then present a fundamental result on the reduced form of stochastic matrices, which is vital in the remaining presentation of this chapter.

A square matrix E is *row stochastic* (*stochastic* for short) if it is nonnegative and every row sum equals 1.

Let $\mathbf{1}$ denote the vector of all 1 components with appropriate dimension. Now we present several well-known properties that stochastic matrices have.

- If a matrix E is stochastic, then $\rho(E) = 1$.

- A nonnegative matrix E is stochastic if and only if $\mathbf{1}$ is an eigenvector of E corresponding to the eigenvalue $\lambda = 1$.

- If matrices E_1 and E_2 are stochastic, then the product $E_1 E_2$ is also stochastic.

Theorem 3.6. *Let E be an $n \times n$ nonnegative matrix. If $\mathcal{G}(E)$ has exactly k closed strong components, then E is cogredient to*

$$\tilde{E} = \begin{pmatrix} E_k & 0 & 0 & 0 \\ 0 & \ddots & 0 & 0 \\ 0 & 0 & E_1 & 0 \\ B_k & \cdots & B_1 & E_0 \end{pmatrix}, \tag{3.5}$$

where E_i, $i = 0, 1, \ldots, k$, are $r_i \times r_i$ matrices and r_i are suitable integers satisfying $0 \le r_0 < n$ and $0 < r_i < n$ ($i = 1, \ldots, k$). If, in addition, E is stochastic, then $\rho(E_i) = 1$ is a simple eigenvalue of $E_i, i = 1, \ldots, k$ and $\rho(E_0) < 1$ when $r_0 \ne 0$.

Proof: If $\mathcal{G}(E)$ has exactly k closed strong components, denote them by

$$\mathcal{G}_1 = (\mathcal{V}_1, \mathcal{E}_1), \ \ldots, \ \mathcal{G}_k = (\mathcal{V}_k, \mathcal{E}_k).$$

If necessary, renumber the nodes and correspondingly permute the rows and columns of E obtaining $\tilde{E}$, such that

$$\mathcal{V}_k = \{1, \ldots, r_k\}$$
$$\mathcal{V}_{k-1} = \{r_k + 1, \ldots, r_k + r_{(k-1)}\}$$
$$\vdots$$
$$\mathcal{V}_1 = \{r_k + \cdots + r_2 + 1, \ldots, r_k + \cdots + r_1\}.$$

Notice that each strong component $\mathcal{G}_i$ is closed by assumption. This means that there are no outgoing arcs leaving $\mathcal{V}_i$. So for any $l \in \mathcal{V}_i$ and any $j \notin \mathcal{V}_i$, the (l, j)-th entry of $\tilde{E}$ is 0. That is, E is cogredient to the matrix in (3.5). When

$$\mathcal{V} = \mathcal{V}_k \cup \cdots \cup \mathcal{V}_1,$$

$r_0 = 0$ and otherwise $r_0 = n - r_1 - \cdots - r_k > 0$.

If, in addition, E is stochastic, then for $i = 1, \ldots, k$, E_i is stochastic and $\rho(E_i) = 1$. Since $\mathcal{G}_i$ is a strong component, it follows from Theorem 3.2 that E_i is irreducible. Then by Theorem 3.5 (Perron-Frobenius theorem) we obtain $\rho(E_i) = 1$ is a simple eigenvalue of E_i for $i = 1, \ldots, k$.

Finally, when $r_0 \neq 0$, denote

$$\mathcal{V}_0 = \mathcal{V} - \mathcal{V}_1 - \cdots - \mathcal{V}_k,$$

which is not empty. We claim that $\mathcal{V}_0$ is not closed in $\mathcal{G}(\tilde{E})$ since otherwise by Lemma 2.1 $\mathcal{G}(\tilde{E})$ has another closed strong component $\mathcal{G}_c = (\mathcal{V}_c, \mathcal{E}_c)$ satisfying $\mathcal{V}_c \subseteq \mathcal{V}_0$, which contradicts that $\mathcal{G}(\tilde{E})$ has exactly k closed strong components. Consequently, there is an arc leading from some node $v_{o_1} \in \mathcal{V}_0$ to some node $v_{i_1} \in \mathcal{V}_i$, which is actually a walk of length of 1.

Next we show that for any integer m there is a walk of length of m from v_{o_1} to some node $v_{i_j} \in \mathcal{V}_i$. If $\mathcal{V}_i$ has only one node v_{i_1}, by recalling that E_i is stochastic, it follows that $E_i = 1$, which in turn implies there is a loop from v_{i_1} to v_{i_1}. Hence there is a walk of length m from v_{o_1} to v_{i_1} by repeatedly passing through the loop. Otherwise if $\mathcal{V}_i$ has more than one node, then there is a walk of any length from v_{i_1} to some other node $v_{i_j} \in \mathcal{V}_i$ since $\mathcal{G}_i$ is strongly connected. Hence for any integer $m \geq 1$ there is a walk of length of m from v_{o_1} to some node $v_{i_j} \in \mathcal{V}_i$. Again, let

$$\mathcal{V}_{o_1} = \mathcal{V}_0 - \{v_{o_1}\}.$$

If $\mathcal{V}_{o_1}$ is not empty then it is also not closed in $\mathcal{G}(E)$ for the same reason that $\mathcal{V}_0$ is not closed. So there is an arc leading from some node $v_{o_2} \in \mathcal{V}_{o_1}$ to some node in

$$\mathcal{V}_1 \cup \cdots \cup \mathcal{V}_k \cup \{v_{o_1}\}.$$

Recalling that some node in $\mathcal{V}_i$ is reachable from v_{o_1}, it follows from the same argument above that for any integer $m \geq 2$ there is a walk of length m from v_{o_2} to some node $v_{j_l} \in \mathcal{V}_j$. Notice that $\mathcal{V}_0$ has at most $n - k$ nodes. Repeating this argument eventually leads to the result that for any integer $m \geq (n - k)$ there is a walk of length of m from any node in $\mathcal{V}_0$ to some node in $\mathcal{V}_1 \cup \cdots \cup \mathcal{V}_k$. Notice that $\tilde{E}^m$ (the m-th power of $\tilde{E}$) will be of the form

$$\tilde{E}^m = \begin{pmatrix} \bar{E}^m & 0 \\ \bar{B}^{(m)} & E_0^m \end{pmatrix},$$

where $\bar{E} = \text{diag}\{E_k, \ldots, E_1\}$ and $\bar{B}^{(m)}$ depends on $\bar{E}, E_0$ and $B_i, i = 1, \ldots, k$. Thus by Theorem 3.3 for any $m \geq (n - k)$ there is at least one entry in each row of $\bar{B}^{(m)}$ that is positive. On the other hand, by the properties of stochastic matrices we know $\tilde{E}^m$ is stochastic. Hence each row sum of E_0^m is less than 1 and therefore $\rho(E_0^m) < 1$ by Theorem 3.1 (Gershgorin disk theorem). Thus the conclusion follows that $\rho(E_0) < 1$.   ■

For example, let

$$E = \begin{pmatrix} 0 & 0 & 1 & 0 \\ 0 & 0 & 0.3 & 0.7 \\ 1 & 0 & 0 & 0 \\ 0 & 0 & 0 & 1 \end{pmatrix}.$$

Then the associated digraph $\mathcal{G}(E)$ is shown in Fig. 3.7 (a). As we can see, this di-

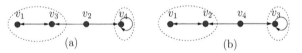

(a)   (b)

Fig. 3.7. Associated digraph with two closed strong components.

graph has two closed strong components, which are circled with dotted lines. By appropriately renumbering the nodes as shown in Fig. 3.7 (b), the corresponding nonnegative matrix obtained from E after the corresponding permutation operation will have the form as (3.5) which is given below:

$$\tilde{E} = PEP^T = \begin{pmatrix} E_2 & 0 & 0 \\ \hline 0 & E_1 & 0 \\ \hline B_2 & B_1 & E_0 \end{pmatrix} = \begin{pmatrix} 0 & 1 & 0 & 0 \\ 1 & 0 & 0 & 0 \\ \hline 0 & 0 & 1 & 0 \\ 0 & 0.3 & 0.7 & 0 \end{pmatrix}.$$

Clearly, $\rho(E_2) = 1$, $\rho(E_1) = 1$, and $\rho(E_0) = 0 < 1$.

3.5 SIA Matrices

Matrix multiplication is associative and the product of two stochastic matrices is again a stochastic matrix. Powers of stochastic matrices will be studied here in detail, with emphasis on convergence. The definition of SIA matrices is introduced by Wolfowitz in [142]. Here we derive a necessary and sufficient condition for a stochastic matrix to be SIA, which will be important in the proof of a main result presented in Chapter 3.

A square matrix E is called *stochastic, indecomposable, and aperiodic (SIA)* if E is stochastic as well as $Q = \lim_{k \to \infty} E^k$ exists and all the rows of Q are the same.

The following theorem gives a graphical characterization of SIA matrices.

Theorem 3.7. *A stochastic matrix E is SIA if and only if $\mathcal{G}(E)$ has a globally reachable node which is aperiodic.*

Proof: ($\Longleftarrow$) If $\mathcal{G}(E)$ has a globally reachable node, then by Theorem 2.1 it has only one closed strong component, say $\mathcal{G}_1 = (\mathcal{V}_1, \mathcal{E}_1)$. We renumber the nodes if necessary such that

$$\mathcal{V}_1 = \{v_1, v_2, \ldots, v_r\}.$$

In other words, there is a permutation matrix P such that

$$\tilde{E} = PEP^T = \begin{pmatrix} E_1 & 0 \\ B_1 & E_0 \end{pmatrix},$$

where E_1 and E_0 are $r \times r$ and $(n-r) \times (n-r)$ matrices respectively. In particular, if $r = n$, then $\tilde{E} = E_1$. Immediately, we know that E_1 is stochastic and so $\rho(E_1) = 1$. Moreover, by the condition that the globally reachable node is aperiodic, it follows from Lemma 3.2 that $\mathcal{G}_1$ is aperiodic since the globally reachable node in $\mathcal{G}(E)$ is also a node in $\mathcal{G}_1$ and $\mathcal{G}_1$ is strongly connected. Then from Theorem 3.4 the matrix E_1, whose associated digraph is $\mathcal{G}_1$, is primitive. By Theorem 3.5 (Perron-Frobenius Theorem), $\rho(E_1) = 1$ is a simple eigenvalue of E_1 and all other eigenvalues of E_1 has modulus less than $\rho(E_1) = 1$. Thus the Jordan form of E_1 is as follows:

$$E_{1JF} = \begin{pmatrix} 1 & 0 \\ 0 & D \end{pmatrix},$$

where $\rho(D) < 1$. Clearly,

$$E_1 = UE_{1JF}U^{-1}$$

and the first column of U is $\mathbf{1}_r$ since $\mathbf{1}$ is an eigenvector of E_1 corresponding to the eigenvalue 1. Denote

$$U = \left(\mathbf{1}_r \, U' \right)$$

and let the first row of U^{-1} be x^T. Then since $\rho(D) < 1$, we obtain

$$\lim_{k \to \infty} E_1^k = U(\lim_{k \to \infty} E_{1JF}^k)U^{-1} = U \begin{pmatrix} 1 & 0 \\ 0 & 0_{(r-1)\times(r-1)} \end{pmatrix} U^{-1} = \mathbf{1}_r \cdot x^T. \qquad (3.6)$$

On the one hand, when $r = n$, it follows from the definition of SIA matrix that $\tilde{E} = E_1$ is SIA and so is E.

On the other hand, when $r < n$, it follows from Theorem 3.6 that $\rho(E_0) < 1$ and so

$$\lim_{k \to \infty} E_0^k = 0. \qquad (3.7)$$

Denote the k-th power of $\tilde{E}$ as follows:

$$\tilde{E}^k = \begin{pmatrix} E_1^k & 0 \\ B_1^{(k)} & E_0^k \end{pmatrix}$$

for some matrix $B_1^{(k)}$ depending upon the matrix E_1, E_0 and B_1. Notice that $\tilde{E}^k$ is still a stochastic matrix for every k. So

$$\left(B_1^{(k)} \, E_0^k \right) \begin{pmatrix} \mathbf{1}_r \\ \mathbf{1}_{(n-r)} \end{pmatrix} = B_1^{(k)} \cdot \mathbf{1}_r + E_0^k \cdot \mathbf{1}_{(n-r)} = \mathbf{1}_{(n-r)},$$

and then we have

$$\begin{aligned} \lim_{k \to \infty} \left(B_1^{(k)} \cdot \mathbf{1}_r \right) &= \lim_{k \to \infty} \left(\mathbf{1}_{(n-r)} - E_0^k \cdot \mathbf{1}_{(n-r)} \right) \\ &= \mathbf{1}_{(n-r)} - \left(\lim_{k \to \infty} E_0^k \right) \cdot \mathbf{1}_{(n-r)} = \mathbf{1}_{(n-r)}. \end{aligned} \qquad (3.8)$$

Express $\tilde{E}^{2k}$ and $\tilde{E}^{2k+1}$ in terms of $\tilde{E}^k$. Then

$$\tilde{E}^{2k} = \tilde{E}^k \cdot \tilde{E}^k = \begin{pmatrix} E_1^k & 0 \\ B_1^{(k)} & E_0^k \end{pmatrix} \begin{pmatrix} E_1^k & 0 \\ B_1^{(k)} & E_0^k \end{pmatrix} = \begin{pmatrix} E_1^{2k} & 0 \\ B_1^{(k)}E_1^k + E_0^k B_1^{(k)} & E_0^{2k} \end{pmatrix}, \qquad (3.9)$$

$$\tilde{E}^{2k+1} = \tilde{E} \cdot \tilde{E}^{2k} = \begin{pmatrix} E_1^{2k+1} & 0 \\ B_1 E_1^{2k} + E_0 \left(B_1^{(k)} E_1^k + E_0^k B_1^{(k)} \right) & E_0^{2k+1} \end{pmatrix}. \qquad (3.10)$$

Considering (3.7) and the fact that each entry of $B_1^{(k)}$ is lower-bounded by zero and upper-bounded by one, we have

$$\lim_{k \to \infty} E_0^k B_1^{(k)} = 0. \qquad (3.11)$$

On the other hand, we have

$$\begin{aligned} B_1^{(k)} E_1^k &= B_1^{(k)} U E_{1JF}^k U^{-1} = B_1^{(k)} \left(\mathbf{1}_r, U' \right) \begin{pmatrix} 1 & 0 \\ 0 & D^k \end{pmatrix} U^{-1} \\ &= \left(B_1^{(k)} \mathbf{1}_r, B_1^{(k)} U' D^k \right) U^{-1}. \end{aligned} \qquad (3.12)$$

Notice that the entries of $B_1^{(k)} U'$ are also bounded and $\lim\limits_{k \to \infty} D^k = 0$. So

$$\lim_{k \to \infty} B_1^{(k)} U' D^k = 0. \qquad (3.13)$$

Combining (3.8) and (3.13), we obtain

$$\begin{aligned} \lim_{k \to \infty} B_1^{(k)} E_1^k &= \left(\lim_{k \to \infty} B_1^{(k)} \mathbf{1}_r, \lim_{k \to \infty} B_1^{(k)} U' D^k \right) U^{-1} \\ &= \left(\mathbf{1}_{(n-r)} \ 0 \right) U^{-1} = \mathbf{1}_{(n-r)} x^T. \end{aligned}$$

Recalling (3.11), then it follows that

$$\begin{aligned} \lim_{k \to \infty} \left(B_1^{(k)} E_1^k + E_0^k B_1^{(k)} \right) &= \lim_{k \to \infty} B_1^{(k)} E_1^k + \lim_{k \to \infty} E_0^k B_1^{(k)} \\ &= \mathbf{1}_{(n-r)} x^T + 0 = \mathbf{1}_{(n-r)} x^T \end{aligned} \qquad (3.14)$$

and therefore from (3.6), (3.7), and (3.14), we have

$$
\lim_{k\to\infty} \tilde{E}^{2k} = \begin{pmatrix} \displaystyle\lim_{k\to\infty} E_1^{2k} & 0 \\ \displaystyle\lim_{k\to\infty}\left(B_1^{(k)}E_1^k + E_0^k B_1^{(k)}\right) & \displaystyle\lim_{k\to\infty} E_0^{2k} \end{pmatrix}
$$

$$
= \begin{pmatrix} \mathbf{1}_r x^T & 0 \\ \mathbf{1}_{(n-r)} x^T & 0 \end{pmatrix} = \begin{pmatrix} \mathbf{1}_n x^T & 0 \end{pmatrix},
$$

$$
\lim_{k\to\infty} \tilde{E}^{2k+1} = \begin{pmatrix} \displaystyle\lim_{k\to\infty} E_1^{2k+1} & 0 \\ \displaystyle\lim_{k\to\infty}\left(B_1 E_1^{2k} + E_0\left(B_1^{(k)}E_1^k + E_0^k B_1^{(k)}\right)\right) & \displaystyle\lim_{k\to\infty} E_0^{2k+1} \end{pmatrix}
$$

$$
= \begin{pmatrix} \mathbf{1}_r x^T & 0 \\ B_1\left(\displaystyle\lim_{k\to\infty} E_1^{2k}\right) + E_0\left(\displaystyle\lim_{k\to\infty}\left(B_1^{(k)}E_1^k + E_0^k B_1^{(k)}\right)\right) & 0 \end{pmatrix}
$$

$$
= \begin{pmatrix} \mathbf{1}_r x^T & 0 \\ B_1 \mathbf{1}_r x^T + E_0 \mathbf{1}_{(n-r)} x^T & 0 \end{pmatrix}
$$

$$
= \begin{pmatrix} \mathbf{1}_r x^T & 0 \\ \mathbf{1}_{(n-r)} x^T & 0 \end{pmatrix} = \begin{pmatrix} \mathbf{1}_n x^T & 0 \end{pmatrix}.
$$

Hence,

$$
\lim_{k\to\infty} \tilde{E}^k = \begin{pmatrix} \mathbf{1}_n x^T & 0 \end{pmatrix},
$$

and so

$$
\lim_{k\to\infty} E^k = T^T \begin{pmatrix} \mathbf{1}_n x^T & 0 \end{pmatrix} T
$$

exists and has identical rows. Then it follows from the definition of SIA matrix that the stochastic matrix E is SIA.

($\Longrightarrow$) If a stochastic matrix E is SIA, we prove by contradiction that $\mathcal{G}(E)$ has a globally reachable node which is aperiodic. Firstly suppose that $\mathcal{G}(E)$ does not have a globally reachable node. Then by Theorem 2.1 it has at least two closed strong components. Thus E is cogredient to

$$
\tilde{E} = \begin{pmatrix} E_2 & 0 & 0 \\ 0 & E_1 & 0 \\ B_2 & B_1 & E_0 \end{pmatrix}.
$$

Recall that E is SIA and so is $\tilde{E}$. That is, $\lim\limits_{k\to\infty} \tilde{E}^k$ exists and has identical rows. However, notice that $\tilde{E}^k$ is of the form

$$\tilde{E}^k = \begin{pmatrix} E_2^k & 0 & 0 \\ 0 & E_1^k & 0 \\ B_2^{(k)} & B_1^{(k)} & E_0^k \end{pmatrix}.$$

Since $\lim\limits_{k\to\infty} \tilde{E}^k$ has identical rows, $\lim\limits_{k\to\infty} \tilde{E}^k$ has to be equal to 0, which contradicts that $\tilde{E}^k$ is stochastic.

Secondly suppose $\mathcal{G}(E)$ has a globally reachable node but this node is d-periodic with $d > 1$. Then by Theorem 2.1 it has only one closed strong component, say $\mathcal{G}_1$, which is d-periodic. Thus E is cogredient to

$$\tilde{E} = \begin{pmatrix} E_1 & 0 \\ B_1 & E_0 \end{pmatrix},$$

where $\mathcal{G}(E_1) = \mathcal{G}_1$. Since $\mathcal{G}(E_1)$ is d-periodic, by Lemma 3.3 we know that the diagonal entries of E_1^k are 0 for any k which is not a multiple of d. Hence, by recalling that $\tilde{E}$ is SIA and noticing the form of $\tilde{E}^k$, we must have

$$\lim_{k\to\infty} \tilde{E}^k = 0.$$

This contradicts that $\tilde{E}^k$ is stochastic. ∎

As an example, let four stochastic matrices be given as follows:

$$E_1 = \begin{pmatrix} 0 & 1 & 0 & 0 \\ 1 & 0 & 0 & 0 \\ 0 & 0.3 & 0 & 0.7 \\ 0 & 0 & 0 & 1 \end{pmatrix} \qquad E_2 = \begin{pmatrix} 0 & 1 & 0 & 0 \\ 0.4 & 0 & 0.6 & 0 \\ 0 & 1 & 0 & 0 \\ 0 & 0 & 1 & 0 \end{pmatrix}$$

$$E_3 = \begin{pmatrix} 0 & 0.5 & 0.5 & 0 \\ 0.2 & 0 & 0.8 & 0 \\ 0 & 1 & 0 & 0 \\ 0 & 0 & 1 & 0 \end{pmatrix} \qquad E_4 = \begin{pmatrix} 0 & 0.9 & 0.1 & 0 \\ 0.5 & 0 & 0.5 & 0 \\ 0 & 0.4 & 0 & 0.6 \\ 0 & 0 & 1 & 0 \end{pmatrix}$$

Then the associated digraphs are given in Fig. 3.8. The digraph $\mathcal{G}(E_1)$ does not have a globally reachable node so E_1 is not SIA. Whereas $\mathcal{G}(E_2)$ has a globally reachable node (the nodes v_1, v_2, and v_3 are all globally reachable node) but

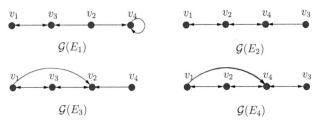

Fig. 3.8. Associated digraphs of SIA matrices.

these nodes are periodic of period 2, so E_2 is not SIA, too. However, $\mathcal{G}(E_3)$ has a globally reachable node v_1 which is aperiodic $(d_1 = \text{g.c.d}\{2, 3\} = 1)$ so E_3 is SIA. Furthermore, $\mathcal{G}(E_4)$ is strongly connected and aperiodic and of course it has a globally reachable node that is aperiodic, so E_4 is SIA. Actually E_4 is a primitive matrix. In fact, all the primitive and stochastic matrices are SIA.

3.6 Wolfowitz Theorem

In a number of important applications the asymptotic behavior of the product of k infinite stochastic matrices as $k \to \infty$ and its dependence on the structure of these matrices are of interest. The theorem stated below, which is due to J. Wolfowitz in 1963, explores this problem. The version we use here is modified from the original version in [142]. The original theorem of Wolfowitz deals with the case with finitely many matrices, whereas here we present a version with possibly infinitely many matrices. Actually, Wolfowitz states this generalization without proof in his paper [142]. For completeness, we provide a proof for this generalization.

For a nonnegative matrix $P = (p_{ij}) \in \mathbb{R}^{n \times n}$, we define $\delta(P)$ by

$$\delta(P) := \max_j \max_{i_1, i_2} |p_{i_1 j} - p_{i_2 j}|.$$

Thus $\delta(P)$ measures, in a certain sense, how different the rows of P are. If the rows of P are identical, $\delta(P) = 0$ and conversely.

On the other hand, we define $\lambda(P)$ by

$$\lambda(P) := 1 - \min_{i_1, i_2, i_1 \neq i_2} \sum_j \min(p_{i_1 j}, p_{i_2 j}). \tag{3.15}$$

If $\lambda(P) < 1$ we will call P a *scrambling matrix*. The condition $\lambda(P) < 1$ implies that, for every pair of rows i_1 and i_2, there exists a column j (which may depend on i_1 and i_2) such that $p_{i_1 j} > 0$ and $p_{i_2 j} > 0$, and conversely. Clearly, whether or not a nonnegative matrix is a scrambling matrix depends solely on its associated digraph. Suppose two nonnegative matrices P_1 and P_2 have the same associated digraph, i.e., $P_1 \sim P_2$. Then if P_1 is a scrambling matrix, so is P_2.

If the matrix P is stochastic, clearly $0 \le \lambda(P) \le 1$. Furthermore, $\lambda(P) = 0$ if and only if $\delta(P) = 0$. This is illustrated next.

Let

$$E_1 = \begin{pmatrix} 0.3 & 0 & 0 & 0.7 \\ 0.2 & 0.2 & 0.5 & 0.1 \\ 0 & 0.2 & 0.8 & 0 \\ 0.6 & 0 & 0.1 & 0.3 \end{pmatrix} \quad \text{and} \quad E_2 = \begin{pmatrix} 0.3 & 0 & 0 & 0.7 \\ 0.2 & 0 & 0.5 & 0.3 \\ 0 & 0.8 & 0 & 0.2 \\ 0.4 & 0.2 & 0.4 & 0 \end{pmatrix}.$$

Then,

$$\delta(E_1) = \max\{0.6, 0.2, 0.8, 0.7\} = 0.8$$
$$\lambda(E_1) = 1 - \min\{0.3, 0, 0.6, 0.7, 0.4, 0.1\} = 1$$

and

$$\delta(E_2) = \max\{0.4, 0.8, 0.5, 0.7\} = 0.8$$
$$\lambda(E_2) = 1 - \min\{0.5, 0.2, 0.3, 0.2, 0.6, 0.2\} = 0.8.$$

Clearly by the definition, E_1 is not a scrambling matrix but E_2 is.

Next we will give several useful lemmas for future reference before Wolfowitz Theorem. Let $\Xi = \{E_1, E_2, \dots\}$ be a finite or infinite set of stochastic matrices of order n. By a product in Ξ of length k we mean the product of k matrices from Ξ (repetitions are permitted). Let m be the number of different associated digraphs of all SIA matrices of order n.

Lemma 3.4 ([142], Lemma 4). *If any product in Ξ is SIA, then all products in Ξ of length $\ge m+1$ are scrambling matrices.*

Lemma 3.5 ([51], Theorem 2). *For any j,*

$$\delta\left(E_{k_1} E_{k_2} \cdots E_{k_j}\right) \le \prod_{i=1}^{j} \lambda\left(E_{k_i}\right).$$

Theorem 3.8 ([142], Wolfowitz, 1963). *If there exists a constant d, $0 \leq d < 1$, such that $\lambda(P) \leq d$ for any product P in Ξ of length $m+1$, then for each infinite sequence, $E_{k_1}, E_{k_2}, \ldots$, there exists a vector x such that*

$$\lim_{j \to \infty} E_{k_j} E_{k_{j-1}} \cdots E_{k_1} = \mathbf{1} x^T.$$

Proof: For any small $\epsilon > 0$, there exists an integer p large enough so that

$$d^p < \epsilon$$

since $0 \leq d < 1$. Let $j^* = (m+1)p$. Then by Lemma 3.5 and the condition that $\lambda(P) \leq d$ for any product P in Ξ of length $m+1$, it follows that, for any integer $j \geq j^*$,

$$\delta\left(E_{k_j} E_{k_{j-1}} \cdots E_{k_1}\right) \leq d^p < \epsilon.$$

Hence, we obtain

$$\lim_{j \to \infty} \delta\left(E_{k_j} E_{k_{j-1}} \cdots E_{k_1}\right) = 0. \tag{3.16}$$

Let

$$C_1 = E_{k_1},$$
$$C_2 = E_{k_2} E_{k_1},$$
$$\vdots$$
$$C_j = E_{k_j} E_{k_{j-1}} \cdots E_{k_1},$$
$$\vdots$$

Denote by c_{rs}^j the (r, s)-th entry of the matrix C_j. From the definition of $\delta(\cdot)$, we obtain that, for any j,

$$0 \leq \max_{r_1, r_2} |c_{r_1 1}^j - c_{r_2 1}^j| \leq \delta(C_j).$$

Combining the inequality above and the equation (3.16) leads to

$$\lim_{j \to \infty} \left(\max_{r_1, r_2} |c_{r_1 1}^j - c_{r_2 1}^j|\right) = 0. \tag{3.17}$$

On the other hand, since each E_{k_j} is stochastic and $C_j = E_{k_j} C_{j-1}$, each entry in the first column of C_j is a convex combination of the entries in the first column of C_{j-1}. So

$$0 \le \max_r \left(c_{r1}^j \right) \le \max_r \left(c_{r1}^{j-1} \right) \quad \text{and} \quad \min_r \left(c_{r1}^{j-1} \right) \le \min_r \left(c_{r1}^j \right) \le 1.$$

Therefore, both $\max_r \left(c_{r1}^j \right)$ and $\min_r \left(c_{r1}^j \right)$ have limits as j tends to ∞, say

$$\lim_{t \to \infty} \max_r \left(c_{r1}^j \right) = a,$$
$$\lim_{t \to \infty} \min_r \left(c_{r1}^j \right) = b.$$

Noticing that

$$\max_r \left(c_{r1}^j \right) - \min_r \left(c_{r1}^j \right) = \max_{r_1, r_2} |c_{r_1 1}^j - c_{r_2}^j|,$$

it follows from (3.17) that $a = b$. Furthermore, notice that, for any j,

$$\min_r \left(c_{r1}^j \right) \le c_{i1}^j \le \max_r \left(c_{r1}^j \right), \quad \forall \, i.$$

Hence, for all $i = 1, \ldots, n$,

$$\lim_{j \to \infty} c_{i1}^j = a = b.$$

This means every entry in the same column of C^j converges to the same constant when j tends to ∞. Therefore, the conclusion follows, that is, there exists a vector x such that

$$\lim_{j \to \infty} E_{k_j} E_{k_{j-1}} \cdots E_{k_1} = \mathbf{1} x^T.$$

∎

4

Generator Matrices and Graphs

Very often problems in biological, physical, and social systems can be reduced to problems involving matrices which, due to certain constraints, have some special structure. One of the most common situations is where the matrix A in question has nonnegative off-diagonal and nonpositive diagonal entries, that is, A is a finite matrix of the type

$$A = \begin{pmatrix} -a_{11} & a_{12} & a_{13} & \cdots \\ a_{21} & -a_{22} & a_{23} & \cdots \\ a_{31} & a_{32} & -a_{33} & \cdots \\ \vdots & \vdots & \vdots & \vdots \end{pmatrix} \qquad (4.1)$$

where the a_{ij} are nonnegative. It should come as no surprise that the theory of nonnegative matrices plays a dominant role in the study of these matrices.

4.1 Metzler Matrices and M-Matrices

A square real matrix A whose off-diagonal entries are nonnegative is called a *Metzler matrix* [86]. These matrices are studied quite often in positive systems. For a linear dynamic system

$$\dot{x}(t) = Ax(t),$$

if A is Metzler, then the transition matrix $\Phi(t) = \exp(At)$ is nonnegative as we will show. This implies that the first quadrant of the state-space is positively invariant.

Closely related properties are clearly possessed by matrices which have the form $-A$, where A is a Metzler matrix.

Matrices of form $B = -A$, where A is a Metzler matrix, may be written in the form

$$B = sI - E,$$

where $s \geq 0$ is a sufficiently large real number to make $E \succeq 0$. If in fact this may be done so that also $s \geq \rho(E)$, then B is called an M-matrix. If further, one can do this so that $s > \rho(E)$ then B is said to be a nonsingular M-matrix.

Examples of Metzler matrix, M-matrix, and nonsingular M-matrix are given as follows, respectively,

$$A_1 = \begin{pmatrix} -2 & 1 & 0 \\ 2 & 0 & 2 \\ 3 & 2 & -1 \end{pmatrix}, \quad A_2 = \begin{pmatrix} 2 & -1 & -1 \\ -2 & 4 & -2 \\ -3 & -2 & 5 \end{pmatrix}, \quad A_3 = \begin{pmatrix} 3 & -1 & -1 \\ -1 & 4 & -2 \\ -2 & -2 & 5 \end{pmatrix}.$$

Now comes a useful result for nonsingular matrices.

Lemma 4.1 ([18], Theorem 6.2.3). *Let A be a Metzler matrix. Then there exists a positive diagonal matrix P such that*

$$AP + PA^T$$

is negative definite if and only if $-A$ is a nonsingular M-matrix.

4.2 Generator Matrices, Laplacians, and Digraphs

In this section we confine ourselves to a special class of Metzler matrices where their row-sums are equal to zero.

A square matrix A is called a *generator matrix* if it is a Metzler matrix with row-sums equal to zero. This notion is originally from continuous time Markov chains (CTMCs).

On the one hand, a generator matrix A could be written as

$$A = -D + E_1, \tag{4.2}$$

where D is a diagonal nonnegative matrix whose diagonal entries are the negative diagonal entries of A and E_1 is a nonnegative matrix with diagonal entries all zero and off-diagonal entries copied from A. Thus, a simple digraph $\mathcal{G}$ can completely capture the structure of a generator matrix, namely, $\mathcal{G}(E_1)$, the associated digraph of the nonnegative matrix E_1. We call this digraph the *associated digraph* of a generator matrix A. In order to avoid causing confusion with the associated digraph of a nonnegative matrix E, we use $\mathcal{G}_A$ to denote the associated digraph of a generator matrix A, whereas we use $\mathcal{G}(E)$ to denote the associated digraph of a nonnegative matrix E, which we once used in the preceding chapter.

On the other hand, a generator matrix A could be also written as

$$A = -sI + E_2, \tag{4.3}$$

where s is a large enough nonnegative scalar so that E_2 is nonnegative. Sometimes we may use this alternative expression for A.

It is worth pointing out that $\mathcal{G}(E_2)$ may be different from $\mathcal{G}_A$, but they surely have the same connectivity properties. For instance, $\mathcal{G}_A$ is strongly connected, quasi-strongly connected, or has k closed strong components if and only if $\mathcal{G}(E_1)$ is strongly connected, quasi-strongly connected, or has k closed strong components. The only difference between these two digraphs is that $\mathcal{G}(E_2)$ may have loops but $\mathcal{G}_A = \mathcal{G}(E_1)$ does not.

As an example, let

$$A = \begin{pmatrix} -2 & 1 & 1 \\ 2 & -3 & 1 \\ 2 & 2 & -4 \end{pmatrix}.$$

The matrix A above is a generator matrix since it is a Metzler matrix and its row sums equal to zero. By the decomposition of (4.2), A is rewritten as

$$A = -D + E_1 = -\begin{pmatrix} 2 & 0 & 0 \\ 0 & 3 & 0 \\ 0 & 0 & 4 \end{pmatrix} + \begin{pmatrix} 0 & 1 & 1 \\ 2 & 0 & 1 \\ 2 & 2 & 0 \end{pmatrix}.$$

By the decomposition of (4.3), A is rewritten as

$$A = -sI + E_2 = -4I + \begin{pmatrix} 2 & 1 & 1 \\ 2 & 1 & 1 \\ 2 & 2 & 0 \end{pmatrix}.$$

The associated digraph $\mathcal{G}_A = \mathcal{G}(E_1)$ and $\mathcal{G}(E_2)$ are given in Fig. 4.1 (a) and (b), respectively, from which we can see that the only difference between them is: $\mathcal{G}_A$ is a simple digraph, but $\mathcal{G}(E_2)$ has several loops.

Fig. 4.1. Associated digraphs of generator matrix.

From the graph theory point of view, a very important matrix associated with a digraph, which is closely related to generator matrices, is the graph Laplacian. Given a digraph $\mathcal{G}$, let D be the diagonal matrix with the out-degree of each node along the diagonal; it is called the *degree matrix* of $\mathcal{G}$. The *Laplacian* of the digraph $\mathcal{G}$ is defined as

$$L = D - E,$$

where E is the adjacency matrix that is defined in the preceding section. It is clear that $-L$ is a generator matrix.

Consider for example a digraph $\mathcal{G}$ in Fig. 4.2. Then the out-degree matrix,

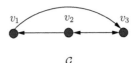

Fig. 4.2. Digraph $\mathcal{G}$.

the adjacency matrix, and the Laplacian are

$$D = \begin{pmatrix} 1 & 0 & 0 \\ 0 & 2 & 0 \\ 0 & 0 & 1 \end{pmatrix}, \quad E = \begin{pmatrix} 0 & 0 & 1 \\ 1 & 0 & 1 \\ 0 & 1 & 0 \end{pmatrix}, \quad \text{and} \quad L = \begin{pmatrix} 1 & 0 & -1 \\ -1 & 2 & -1 \\ 0 & -1 & 1 \end{pmatrix},$$

respectively.

4.3 The Exponential of Generator Matrices

Consider the system

$$\dot{x} = Ax.$$

The well-known solution is

$$x(t) = \exp(At)x(0).$$

The matrix $\exp(At)$ is usually called the *transition matrix* (the exponential). Now we present a property of the exponential of generator matrices.

Theorem 4.1. *If A is a generator matrix, then $\exp(At)$ is a stochastic matrix for each $t > 0$.*

Proof: First, we show that, for each $t > 0$, $\exp(At)$ is a nonnegative matrix. From (4.3), it follows that for each $t > 0$,

$$\exp(At) = \exp(-stI + E_2 t) = e^{-st} \exp(E_2 t) = e^{-st} \cdot \left(I + E_2 t + \frac{(E_2 t)^2}{2!} + \cdots \right).$$

It can be easily verified that any integer power of the nonnegative matrix E_2 is also nonnegative. Hence, looking at the right-hand side of the equality above, we see that $\exp(At)$ is nonnegative for each $t > 0$.

For the exponential to be stochastic, it requires in addition that the rows sum to one. The matrix A has row sums equal to zero. That is, $A \cdot \mathbf{1} = 0$. Hence we obtain

$$\exp(At)\mathbf{1} = \left(I + At + \frac{(At)^2}{2!} + \cdots \right) \mathbf{1} = \mathbf{1} + 0 + 0 + \cdots = \mathbf{1}.$$

Therefore, from the power expansion above we see that $\exp(At)$ is stochastic for each $t > 0$. ∎

4.4 Algebraic Properties of Generator Matrices

We now turn to study the eigenvalues of a generator matrix with a very nice graphical explanation.

Throughout this section, we let A be a generator matrix of order n and let $\mathcal{G}_A$ denote the associated digraph corresponding to A.

Theorem 4.2. *Zero is an eigenvalue of A and $\mathbf{1}$ is an associated right-eigenvector, while all other eigenvalues have negative real part.*

Proof: By the definition of generator matrix, it follows that $A\mathbf{1} = 0$. So 0 is its eigenvalue and $\mathbf{1}$ is its associated right-eigenvector. Furthermore, it follows from

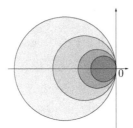

Fig. 4.3. Gershgorin discs of generator matrix.

Theorem 3.1 (Gershgorin disk theorem) that all other eigenvalues have negative real part (see Fig. 4.3). ∎

Theorem 4.3. *If $\mathcal{G}_A$ is strongly connected, the zero eigenvalue of A is simple. Furthermore, associated with the zero eigenvalue, a right-eigenvector is $\mathbf{1}$ and a left-eigenvector is a positive vector.*

Proof: The generator matrix A may be written as

$$A = -sI + E,$$

where s is a positive scalar large enough so that E is nonnegative. Since A is a generator matrix, all rows of A sum to zero. So all rows of E sum to s and the matrix E/s is stochastic. Thus by the properties of stochastic matrices, $\rho(E/s) = 1$ and therefore $\rho(E) = s$.

On the other hand, if $\mathcal{G}_A$ is strongly connected then $\mathcal{G}(E)$ is also strongly connected. Therefore E is irreducible by Theorem 3.2 and $\rho(E) = s$ is a simple eigenvalue of E by Theorem 3.5 (Perron-Frobenius theorem). Furthermore, E has a positive right-eigenvector and a positive left-eigenvector corresponding to the eigenvalue $\rho(E) = s$. Clearly this right-eigenvector is $\mathbf{1}$. In addition, we denote the positive left-eigenvector by x^T. Notice that

$$A = -sI + E,$$

so 0 is its simple eigenvalue and its associated right-eigenvector and left-eigenvector are $\mathbf{1}$ and x^T, respectively. ∎

Theorem 4.4. *For the zero eigenvalue of A, algebraic multiplicity equals geometric multiplicity and equals the number of closed strong components in $\mathcal{G}_A$.*

Proof: Write the generator matrix A as

$$A = -sI + E,$$

where s is a positive scalar and E is a nonnegative matrix. Furthermore, E/s is stochastic. Without loss of generality, say $\mathcal{G}_A$ has exactly k closed strong components. Then so does $\mathcal{G}(E)$. By Theorem 3.6, the nonnegative matrix E is cogredient to

$$\tilde{E} = \begin{pmatrix} E_k & 0 & 0 & 0 \\ 0 & \ddots & 0 & 0 \\ 0 & 0 & E_1 & 0 \\ F_k & \cdots & F_1 & E_0 \end{pmatrix},$$

where each block in the matrix form above has a suitable dimension and only E_0 could have possibly zero dimension. Notice that $\tilde{E}$ is obtained from E only by permutation operation, so $\tilde{E}/s$ is also stochastic. Then by Theorem 3.6 again,

$$\rho\left(E_i/s\right) = 1$$

is a simple eigenvalue of E_i/s for $i = 1, \ldots, k$ and

$$\rho\left(E_0/s\right) < 1$$

if the dimension of E_0 is not zero. Moreover, $\mathbf{1}$ with suitable dimension is the eigenvector of each block E_i/s for $i = 1, \ldots, k$ corresponding to the eigenvalue $\rho\left(E_i/s\right) = 1$. Hence, $\tilde{E}$ has the property that s is its eigenvalue of algebraic and geometric multiplicity k. This in turn implies that E has the same property. By

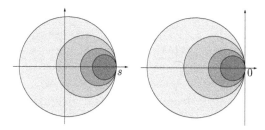

Fig. 4.4. Shifting s units.

shifting s units left, we obtain that A has a zero eigenvalue of algebraic and geometric multiplicity k (see Fig. 4.4).   ■

4.5 $H(\alpha, m)$ Stability

Now we relate the structure of generator matrices to a special type of stability that we call $H(\alpha, m)$ stability, which is modified from [54] in order to better suit our applications.

Given the matrices $A \in \mathbb{R}^{n \times m}$ (with elements $A = (a_{ij})$) and $B \in \mathbb{R}^{p \times q}$, the *Kronecker product* of A and B, denoted $A \otimes B$, is the $np \times mq$ matrix

$$A \otimes B = \begin{pmatrix} a_{11}B & \cdots & a_{1m}B \\ \vdots & \ddots & \vdots \\ a_{n1}B & \cdots & a_{nm}B \end{pmatrix} \in \mathbb{R}^{np \times mq}.$$

The Kronecker product has several useful properties. If the orders of the matrices involved are such that all the operations below are defined, then

(a) $(A + B) \otimes C = A \otimes C + B \otimes C$;

(b) $(A \otimes B)^T = A^T \otimes B^T$;

(c) If $A, C \in \mathbb{R}^{m \times m}$ and $B, D \in \mathbb{R}^{n \times n}$, then $(A \otimes B)(C \otimes D) = (AC) \otimes (BD)$;

(d) If $\lambda_1, \ldots, \lambda_m$ are the eigenvalues of $A \in \mathbb{R}^{m \times m}$ and $\mu_1, \ldots, \mu_n$ are the eigenvalues of $B \in \mathbb{R}^{n \times n}$, then the eigenvalues of $A \otimes B$ are the mn numbers $\lambda_r \mu_s$, $r = 1, \ldots, m$, $s = 1, \ldots n$.

Let

$$\alpha = \{\alpha_1, \alpha_2, \ldots, \alpha_p\}$$

be a partition of $\{1, 2, \ldots, n\}$. A block diagonal matrix with diagonal blocks indexed by $\alpha_1, \alpha_2, \ldots, \alpha_p$ is said to be α-*diagonal*. A scalar multiple of the identity matrix is said to be a *scalar matrix*. An α-diagonal matrix D is said to be an α-*scalar matrix* if each block matrix in D is a scalar matrix.

For $\alpha = \{\{1, 2\}, \{3\}\}$, simple examples of α-diagonal matrix and α-scalar matrix are, respectively,

$$D_1 = \begin{pmatrix} 1 & 3 & 0 \\ 2 & 2 & 0 \\ 0 & 0 & 5 \end{pmatrix} \quad \text{and} \quad D_2 = \begin{pmatrix} 2 & 0 & 0 \\ 0 & 2 & 0 \\ 0 & 0 & 5 \end{pmatrix}.$$

Definition 4.1. *Let $\alpha = \{\alpha_1, \alpha_2, \ldots, \alpha_p\}$ be a partition of $\{1, 2, \ldots, n\}$ and $m \geq 0$ be an integer. An $n \times n$ matrix A is said to be $H(\alpha, m)$-stable if, for every α-diagonal symmetric positive definite matrix S, zero is an eigenvalue of SA of algebraic and geometric multiplicity m, while all other eigenvalues have negative real part.*

Obviously, an $H(\alpha, 0)$-stable matrix is asymptotically stable (Hurwitz), but the converse does not hold in general. A simple counterexample is

$$A = \begin{pmatrix} -5 & -4 \\ 4 & 3 \end{pmatrix} \quad \text{and} \quad \alpha = \{\{1\}, \{2\}\}\,.$$

Then A is asymptotically stable with eigenvalues $\{-1, -1\}$ but SA is unstable with the choice of α-diagonal symmetric positive definite matrix

$$S = \begin{pmatrix} 1 & 0 \\ 0 & 2 \end{pmatrix}\,.$$

In what follows, we let A be a generator matrix of order n and let $\mathcal{G}_A$ denote the associated digraph corresponding to A. Moreover, we let $A_{(m)}$ be the Kronecker product of A and I_m, an $m \times m$ identity matrix, i.e., $A_{(m)} = A \otimes I_m$, and let

$$\alpha = \{\{1, \ldots, m\}, \{m+1, \ldots, 2m\}, \ldots, \{(n-1)m+1, nm\}\}$$

be a partition of $\{1, 2, \ldots, nm\}$.

Theorem 4.5. *If $\mathcal{G}_A$ is strongly connected then $A_{(m)}$ is $H(\alpha, m)$ stable.*

Proof: If $\mathcal{G}_A$ is strongly connected, then by Theorem 4.3, zero is a simple eigenvalue of A with an associated right-eigenvector $\mathbf{1}$ and a positive left eigenvector $x^T = (x_1 \cdots x_n)$. That is,

$$A\mathbf{1} = 0 \quad \text{and} \quad x^T A = 0.$$

Furthermore,

$$\ker(A) = \text{span}\{\mathbf{1}\}.$$

Let $P = \text{diag}(x_1, x_2, \ldots, x_n)$. Thus, P is positive definite and $P\mathbf{1} = x$. Let

$$Q = A^T P + PA.$$

Then we have

$$Q\mathbf{1} = A^T P\mathbf{1} + PA\mathbf{1} = A^T x = 0. \tag{4.4}$$

Notice that a generator matrix A may be written as

$$A = -D + E,$$

where D is a nonnegative diagonal matrix whose diagonal entries are the negative diagonal entries in A, and E is a nonnegative matrix whose diagonal entries are zero and whose off-diagonal entries are the same as the off-diagonal entries in A. Thus,

$$Q = A^T P + PA = (-D + E)^T P + P(-D + E) = -2PD + (E^T P + PE).$$

Recalling that P is a positive diagonal matrix, we know that PD is also a positive diagonal matrix and $E^T P + PE$ is nonnegative. This implies that Q has nonnegative off-diagonal entries. Combining with (4.4), it follows that Q is also a generator matrix. Moreover, recall that $\mathcal{G}_A$ is just the associated digraph $\mathcal{G}(E)$. So $\mathcal{G}(E)$ is strongly connected. Since P is a positive diagonal matrix, we obtain

$$\mathcal{G}(PE) = \mathcal{G}(E).$$

In addition,

$$\mathcal{G}(E^T P + PE) = \mathcal{G}(PE) \cup \mathcal{G}(E^T P) = \mathcal{G}(E) \cup \mathcal{G}(E^T P).$$

Hence, $\mathcal{G}\left(E^T P + PE\right)$ is strongly connected and so is $\mathcal{G}_Q$. Applying Theorem 4.3 again, we obtain

$$\ker(Q) = \mathrm{span}\{\mathbf{1}\}.$$

Also, by Theorem 4.2, except for the zero eigenvalue, the other eigenvalues of Q are in the open left half-plane, implying that Q is negative semi-definite.

We now define a positive-definite α-scalar matrix $P_{(m)} = P \otimes I_m$ and, then by properties (a), (b) and (c) of the Kronecker product, we have

$$
\begin{aligned}
A_{(m)}^T P_{(m)} + P_{(m)} A_{(m)} &= (A \otimes I_m)^T (P \otimes I_m) + (P \otimes I_m)(A \otimes I_m) \\
&= (A^T P) \otimes I_m + (PA) \otimes I_m \\
&= (A^T P + PA) \otimes I_m \\
&= Q \otimes I_m \\
&= Q_{(m)}.
\end{aligned}
$$

Using property (d) of the Kronecker product, we conclude that both $A_{(m)}$ and $Q_{(m)}$ have zero eigenvalue of algebraic and geometric multiplicity m and the others are in the open left half-plane. Hence,

$$
\ker\left(Q_{(m)}\right) = \ker\left(A_{(m)}\right)
$$

and $Q_{(m)}$ is negative semi-definite.

Given any α-diagonal symmetric positive definite matrix S, its inverse S^{-1} is also α-diagonal, symmetric, and positive definite. Recall that $P_{(m)}$ is a positive definite α-scalar matrix, so we obtain that

$$
P_{(m)} S^{-1} = S^{-1} P_{(m)},
$$

which is positive definite. It now follows that

$$
\left(SA_{(m)}\right)^T \left(S^{-1} P_{(m)}\right) + \left(P_{(m)} S^{-1}\right) \left(SA_{(m)}\right) = A_{(m)}^T P_{(m)} + P_{(m)} A_{(m)} = Q_{(m)}.
$$

Thus, by the well-known Lyapunov theory, the eigenvalues of $SA_{(m)}$ are in the open left half-plane or on the imaginary axis. Further, from the spectral properties of $A_{(m)}$ we can infer that zero is an eigenvalue of $SA_{(m)}$ of algebraic and geometric multiplicity m. Therefore, to show $A_{(m)}$ is $H(\alpha, m)$ stable, it remains to show that no other eigenvalues of $SA_{(m)}$, except these m zero eigenvalues, are on the imaginary axis. Suppose by way of contradiction that $\lambda = j\omega$ ($\omega \neq 0$) is one of the eigenvalues with corresponding eigenvector y. Then

$$
\begin{aligned}
0 &= y^* \left\{ \left(SA_{(m)}\right)^* \left(S^{-1} P_{(m)}\right) + \left(P_{(m)} S^{-1}\right) \left(SA_{(m)}\right) - Q_{(m)} \right\} y \\
&= -j\omega y^* \left(S^{-1} P_{(m)}\right) y + j\omega y^* \left(P_{(m)} S^{-1}\right) y - y^* Q_{(m)} y \\
&= -y^* Q_{(m)} y.
\end{aligned}
$$

It follows that $y \in \ker\left(Q_{(m)}\right)$ and therefore $\omega = 0$, a contradiction.   ∎

Theorem 4.6. *The graph $\mathcal{G}_A$ has a globally reachable node if and only if $A_{(m)}$ is $H(\alpha, m)$ stable.*

Proof: ($\Longrightarrow$) If $\mathcal{G}_A$ has a globally reachable node then by Theorem 2.1 it has only one closed strong component, say $\mathcal{G}_1 = (\mathcal{V}_1, \mathcal{E}_1)$. If $\mathcal{G}_1 = \mathcal{G}_A$, it means $\mathcal{G}_A$ is strongly connected and then by Theorem 4.5 $A_{(m)}$ is $H(\alpha, m)$ stable. Otherwise, we let

$$\mathcal{V}_1 = \{1, 2, \ldots, r\}$$

(without loss of generality since if necessary we could renumber the nodes) where $r < n$. Recall that A can be written as

$$A = -sI + E,$$

where s and E are positive scalar and nonnegative matrix, respectively. The graph $\mathcal{G}_A$ has only one closed strong component and so does $\mathcal{G}(E)$. Then by Theorem 3.6, E is of the form

$$E = \begin{pmatrix} E_1 & 0 \\ B_1 & E_0 \end{pmatrix},$$

where E_1 is $r \times r$ matrix and $\rho(E_0) < s$. Similarly, A is of the form

$$A = \begin{pmatrix} A_1 & 0 \\ B_1 & A_0 \end{pmatrix} = \begin{pmatrix} -sI + E_1 & 0 \\ B_1 & -sI + E_0 \end{pmatrix}.$$

Notice that $\mathcal{G}_{A_1}$ is just $\mathcal{G}_1$. So it is strongly connected. Then it follows from Theorem 4.5 that A_1 is $H(\alpha_1, m)$ stable where

$$\alpha_1 = \{\{1, \ldots, m\}, \ldots, \{(r-1)m+1, \ldots, rm\}\}.$$

On the other hand, notice that

$$A_0 = -sI + E_0$$

and that

$$\rho(E_0) < s.$$

So it follows from the definition of nonsingular M-matrix that $-A_0$ is a nonsingular M-matrix. Thus by Lemma 4.1 there exists a positive diagonal matrix P such that

$$Q = A_0^T P + PA_0$$

is negative definite. Applying the Kronecker product with I_m to both sides of the above equation and using properties (a), (b) and (c) of the Kronecker product yield

$$Q_{(m)} = A_{0_{(m)}}^T P_{(m)} + P_{(m)} A_{0_{(m)}}.$$

Clearly, $P_{(m)}$ is an α-scalar positive definite matrix and $Q_{(m)}$ is negative definite.

Let

$$\alpha_2 = \{\{1, \ldots, m\}, \ldots, \{(n - r - 1)m + 1, \ldots, (n - r)m\}\}.$$

Then for any α_2-diagonal symmetric positive definite matrix S_2,

$$P_{(m)} S_2^{-1} = S_2^{-1} P_{(m)},$$

which is positive definite. It can be easily verified that

$$\left(S_2 A_{0_{(m)}}\right)^T \left(S_2^{-1} P_{(m)}\right) + \left(P_{(m)} S_2^{-1}\right)\left(S_2 A_{0_{(m)}}\right) = Q_{(m)}.$$

Thus by the well-known Lyapunov theorem, all the eigenvalues of $S_2 A_{0_{(m)}}$ have negative real part and therefore $A_{0_{(m)}}$ is $H(\alpha_2, 0)$ stable.

Finally notice that, for any α-diagonal positive definite symmetric matrix S, it can be written as

$$S = \begin{pmatrix} S_1 & 0 \\ 0 & S_2 \end{pmatrix}$$

with suitable dimension for each block. Furthermore,

$$SA_{(m)} = \begin{pmatrix} S_1 A_{1_{(m)}} & 0 \\ S_2 B_{1_{(m)}} & S_2 A_{0_{(m)}} \end{pmatrix}.$$

Hence, it follows from the argument above and the definition of $H(\alpha, m)$ stability that $A_{(m)}$ is $H(\alpha, m)$ stable.

($\Longleftarrow$) On the other hand, if $A_{(m)}$ is $H(\alpha, m)$ stable then $A_{(m)}$ has a zero eigenvalue of algebraic multiplicity m by its definition. Applying the property (d) of Kronecker product leads to that A has a simple eigenvalue at zero. By Theorem 4.4 and Theorem 2.1, the digraph $\mathcal{G}_A$ has a globally reachable node. ∎

Part III

Analysis of Coupled Cell Systems

5

Linear Coupled Cell Systems

We study in this chapter the stability property and the agreement problem of linear coupled cell systems with fixed and dynamic topologies. Such systems are represented by switched continuous time linear systems whose system matrices are generator matrices. The equilibrium set of the system contains all states with identical state components. State agreement implies that all state components converge to a common value. This class of problems naturally arise in the context of distributed decision making problems, coordination and consensus seeking problems in multi-agent systems, and synchronization problems.

For reasons of clarity, we deliberately restrict attention to coupled systems with linear dynamics throughout the present chapter. Indeed, all the linear results presented in this chapter can be obtained from the more general nonlinear results in the next chapter. But we use completely different techniques to prove them, mainly based on the matrix theory developed in the preceding chapters. Also, it is the first step in studying such coupled cell systems. So we keep it in the monograph as a self-contained chapter.

5.1 State Model and Interaction Graph

Very often problems in biological, physical, social systems, and in coordination control of multi-agent systems can be reduced to problems involving generator matrices. This important type of matrices was originally studied in stochastic processes, in particular, Markov chains, of continuous time setup. It is of recent interest in the field of linear coupled cell systems, especially with applications in coordination and control for multiple agents.

Consider a coupled cell system with n cells that interact each other with the linear coupling:

$$\dot{x}_i = \sum_{j=1}^{n} a_{ij}(p)\,(x_j - x_i),\quad i = 1, 2, \ldots, n \tag{5.1}$$

where $x_i \in \mathbb{R}^m$ is the state of *cell* or *agent* i, the index p lives in a finite set $\mathcal{P}$. Additionally, the coefficients $a_{ij}(p) \geq 0$ for all i, j and p. When $a_{ij}(p) > 0$ for some i, j, p, it means the dynamics of x_i is influenced by the state x_j.

Introducing the *aggregate state* $x = (x_1, \ldots, x_n) \in \mathbb{R}^{mn}$, the linear coupled cells system can be written in a matrix form:

$$\dot{x} = \big(A_p \otimes I_m\big)x,\quad p \in \mathcal{P} \tag{5.2}$$

where I_m is the identity matrix of order m, and for each $p \in \mathcal{P}$, the matrix A_p is defined in the following way: the ijth off-diagonal entry is $a_{ij}(p)$; the ith diagonal entry is $-\sum_{j=1}^{n} a_{ij}(p)$. Clearly, by the definition, all the matrices A_p, $p \in \mathcal{P}$, are generator matrices. Each individual component model $\dot{x} = \big(A_p \otimes I_m\big)x$ for $p \in \mathcal{P}$ is called a *mode* of the family (5.2).

To define a switched coupled cell system generated by the above family, we need the notion of a *switching signal*. This is a piecewise constant function $\sigma : \mathbb{R} \to \mathcal{P}$. Such a function σ has a finite number of discontinuities at the *switching times* on every bounded time interval and takes a constant value on every interval between two consecutive switching times. The role of σ is to specify, at each time instant t, the index $\sigma(t) \in \mathcal{P}$ of the *active mode*, i.e., the system from the family (5.2) that is currently being followed. In some applications, the switching signal $\sigma(t)$ may be known a priori, but in some other applications, it may be unknown and will be specified only when we come to specific examples. Thus, a *switched linear coupled cell system* is described by the equation

$$\dot{x}(t) = \big(A_{\sigma(t)} \otimes I_m\big)x(t). \tag{5.3}$$

For the coupled cell system above, we can use a dynamic graph called *interaction graph* to capture the coupling structure of the n cells.

Definition 5.1. *The* interaction graph $\mathcal{G}(t) = (\mathcal{V}, \mathcal{E}(t))$ *consists of*

- *a finite set $\mathcal{V}$ of n nodes, each node i modeling cell i;*
- *a (time-varying) arc set $\mathcal{E}(t)$ representing the links between cells at time t. An arc from node j to node i at t indicates that cell j is a neighbor of cell i in the sense that $a_{ij}(\sigma(t)) \neq 0$. The set of neighbors of cell i at t is denoted by $\mathcal{N}_i(t)$.*

By this definition, the interaction graph might be fixed when $\sigma(t)$ is a constant, or time-varying otherwise. Corresponding to a switching signal $\sigma(t)$, the interaction graph $\mathcal{G}(t)$ actually switches among a family of digraphs $\{\mathcal{G}_p : p \in \mathcal{P}\}$, where $\mathcal{G}_p$ is called the *mode graph* at p defined as follows:

$$\mathcal{G}_p := \mathcal{G}(t) \text{ when } \sigma(t) \equiv p.$$

It is worth pointing out that each mode graph $\mathcal{G}_p$ is the opposite digraph of the associated digraph $\mathcal{G}_{A_p}$ of the generator matrix A_p.

The next example combines several of the concepts presented thus far.

Consider a linear coupled system of three cells labelled 1, 2, and 3. Let $x_1, x_2, x_3 \in \mathbb{R}^2$ be their states respectively. Suppose there are three possible modes in the family (i.e., $\mathcal{P} = \{1, 2, 3\}$):

$$
p = 1: \qquad\qquad p = 2: \qquad\qquad p = 3:
$$

$$
\left\{
\begin{array}{l}
\dot{x}_1 = 3(x_2 - x_1) \\
\dot{x}_2 = 2(x_3 - x_2) + (x_1 - x_2) \\
\dot{x}_3 = 4(x_1 - x_3)
\end{array}
\right\},
\left\{
\begin{array}{l}
\dot{x}_1 = 3(x_3 - x_1) \\
\dot{x}_2 = 0 \\
\dot{x}_3 = 4(x_2 - x_3)
\end{array}
\right\},
\left\{
\begin{array}{l}
\dot{x}_1 = x_3 - x_1 \\
\dot{x}_2 = 0 \\
\dot{x}_3 = x_2 - x_3
\end{array}
\right\}.
$$

The corresponding matrix form for the mode $p = 1$ is

$$
\begin{pmatrix} \dot{x}_1 \\ \dot{x}_2 \\ \dot{x}_3 \end{pmatrix} =
\left(
\begin{array}{c|c|c}
-3I_2 & 3I_2 & 0 \\ \hline
I_2 & -3I_2 & 2I_2 \\ \hline
4I_2 & 0 & -4I_2
\end{array}
\right)
\begin{pmatrix} x_1 \\ x_2 \\ x_3 \end{pmatrix} =
\left(
\begin{pmatrix}
-3 & 3 & 0 \\
1 & -3 & 2 \\
4 & 0 & -4
\end{pmatrix}
\otimes I_2
\right)
\begin{pmatrix} x_1 \\ x_2 \\ x_3 \end{pmatrix}.
$$

Corresponding to the three modes, the mode graphs $\mathcal{G}_1$, $\mathcal{G}_2$, and $\mathcal{G}_3$, are depicted in Fig. 5.1. Further, suppose we are given a switched signal $\sigma(t)$ shown in Fig. 5.2.

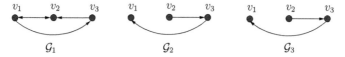

Fig. 5.1. Mode graphs.

This gives rise to a switched linear coupled cell system of the form (5.3) and the interaction graph $\mathcal{G}(t)$ that switches among $\{\mathcal{G}_1, \mathcal{G}_2, \mathcal{G}_3\}$.

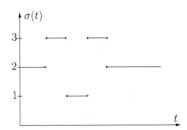

Fig. 5.2. Switching signal $\sigma(t)$.

Having introduced the state model for linear coupled cell systems and the interaction graph representing the coupling topology of cells, we are now ready to give some formal definitions for the state agreement problem we will deal with in the chapter.

Let us first define the subspace Ω as

$$\Omega = \big\{x \in \mathbb{R}^{mn} : x_1 = x_2 = \cdots = x_n\big\}.$$

Notice that for every $p \in \mathcal{P}$, the matrix A_p is a generator matrix. This implies that for every switching signal $\sigma(t)$

$$\big(A_{\sigma(t)} \otimes I_m\big)\bar{x} = 0 \quad \text{for all } \bar{x} \in \Omega.$$

Hence, the subspace Ω is an equilibria set. In this monograph, we call this particular $\bar{x} \in \Omega$ an *agreement state*. Let x^0 be any initial state in $\mathbb{R}^{mn}$ and let $x(t)$ be the solution of the coupled cell system (5.3) starting at (t_0, x^0) (i.e., $x(t_0) = x_0$).

Definition 5.2. *The system (5.3) is said to* achieve (state) agreement *if for any* $x^0 \in \mathbb{R}^{mn}$ *there exists an* $\bar{x} \in \Omega$ *such that*

$$\lim_{t \to \infty} x(t) = \bar{x}.$$

Definition 5.3. *The system (5.3) is said to* achieve (state) agreement uniformly *if for any* $c > 0$ *and for any* $\|x^0\| \le c$ *there exists an* $\bar{x} \in \Omega$ *such that*

$$\lim_{t \to \infty} x(t) = \bar{x} \ \text{uniformly in } t_0, x^0.$$

Note that the definition above is equivalent to the following statement: For any $c > 0$ and for any $\varepsilon > 0$ there exists a $T > 0$ such that for any t_0 and for any $\|x^0\| \le c$,

$$(\exists \bar{x} \in \Omega)(\forall t \ge t_0 + T) \ \ \|x(t) - \bar{x}\| \le \varepsilon.$$

As we know, there is no distinction between stability and uniform stability for autonomous systems. Likewise, there is no distinction between achieving state agreement and achieving state agreement uniformly for autonomous systems. In addition, it is worth pointing out that stability together with agreement for our systems implies: every equilibrium in Ω is semistable (see Appendix A).

We will investigate the stability property of agreement state and study necessary and sufficient conditions so that the linear coupled cell system achieves state agreement (uniformly).

When $\sigma(t)$ is a constant, that is, $\sigma(t) \equiv p$ for some $p \in \mathcal{P}$, then the switched system becomes time-invariant. We say the coupled cell system has a fixed topology. For simplicity, we drop the subscript for the coupled cell system with fixed topology. Thus, we have the following matrix form:

$$\dot{x} = \big(A \otimes I_m\big)x. \tag{5.4}$$

Correspondingly, we will just use $\mathcal{G}$ rather than $\mathcal{G}(t)$ to denote the interaction graph for coupled cell system with fixed topology. In what follows we will start with the coupled cell system with fixed topology.

5.2 Fixed Topology: Cyclic Coupling

In this section our analysis focuses on a particular fixed topology called *cyclic coupling structure*.

Consider a system of n cells with cyclic coupling structure,

$$\dot{x}_1 = x_2 - x_1$$
$$\dot{x}_2 = x_3 - x_2$$
$$\vdots \tag{5.5}$$
$$\dot{x}_n = x_1 - x_n$$

where $x_i \in \mathbb{R}^m$ is the state of cell i. As we will see later, the case of $m > 1$ is a trivial extension of the case of $m = 1$. So we now assume $m = 1$ for simplicity.

Thus the system above in matrix form is given by

$$\dot{x} = Ax,$$

where A has the form

$$A = P - I$$

and P is the *permutation matrix* obtained by taking I and putting its first row at the bottom:

$$P = \begin{pmatrix} 0 \ 1 \ 0 \ldots \ 0 \\ 0 \ 0 \ 1 \ldots \ 0 \\ \vdots \ \vdots \ \vdots \ \ \vdots \ \ \vdots \\ 1 \ 0 \ 0 \ldots \ 0 \end{pmatrix}. \tag{5.6}$$

As an example, the interaction graph of six cells with such a coupling structure is given in Fig. 5.3.

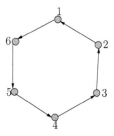

Fig. 5.3. Cyclic coupling structure.

Let x^0 be any initial state for the system (5.5). Then we have the following result.

Theorem 5.1. *The linear coupled cell system (5.5) achieves state agreement. Furthermore,*

$$\lim_{t \to \infty} x(t) = \left(\frac{x_1^0 + x_2^0 + \cdots + x_n^0}{n} \right) \mathbf{1}.$$

Proof: It is easy to obtain that the characteristic polynomial of P is $s^n - 1$. So the eigenvalues of P are the nth roots of unity, and therefore the eigenvalues of A are these roots of unity shifted left by 1. That is, A has an eigenvalue at the origin and $n - 1$ distinct eigenvalues strictly in the left half-plane. Moreover, it can be easily seen that $\ker(A) = \Omega$. Hence, for any initial state, the trajectory asymptotically converges to Ω.

In addition, one sees

$$(\dot{x}_1 + \cdots + \dot{x}_n)/n = (\mathbf{1}^T Ax)/n = 0.$$

So

$$\lim_{t \to \infty} x(t) = \left(\frac{x_1^0 + x_2^0 + \cdots + x_n^0}{n} \right) \mathbf{1}.$$

Thus, the system achieves state agreement.   ∎

5.3 Fixed Topology: Arbitrary Coupling

This section we turn to the general coupling structure of n cells. But again, it is assumed that the coupling topology is fixed.

Consider the following linear coupled cell system represented by equations of the form

$$\begin{cases} \dot{x}_1 = \sum_{j=1}^n a_{1j}(x_j - x_1) \\ \qquad \vdots \\ \dot{x}_n = \sum_{j=1}^n a_{nj}(x_j - x_n) \end{cases} \tag{5.7}$$

where $x_i \in \mathbb{R}^m$ and $a_{ij} \geq 0$. Let x^0 be an initial state. Define the interaction graph $\mathcal{G}$ of the coupled cell system (5.7) in the way of Definition 5.1.

Firstly, we assume $m = 1$, so the system (5.7) in matrix form is given by

$$\dot{x} = Ax, \tag{5.8}$$

where A is a generator matrix.

Theorem 5.2. *Every agreement state is stable for the linear coupled cell system (5.7).*

The proof is similar to the one for Theorem 5.5 to be given later. So we omit it here.

Theorem 5.3. *The linear coupled cell system (5.8) achieves state agreement if and only if the interaction graph $\mathcal{G}$ is quasi-strongly connected. Moreover,*

$$\lim_{t \to \infty} x(t) = \left(\frac{c^T x^0}{c^T \mathbf{1}} \right) \mathbf{1},$$

where c^T is a left-eigenvector of A corresponding to the zero eigenvalue.

Proof: ($\Longleftarrow$) If the interaction graph $\mathcal{G}$ is quasi-strongly connected, it follows from Theorem 2.1 and Theorem 4.4 that A has one zero eigenvalue and all other eigenvalues have negative real part. By inspection, an associated eigenvector is $\mathbf{1}$ and so $\ker(A) = \Omega$. This implies that the trajectories asymptotically converge to Ω.

Let c^T be a left eigenvector of A corresponding to the zero eigenvalue. Define a new variable $y = c^T x$. Then we have

$$\dot{y} = c^T \dot{x} = c^T A x = 0,$$

which implies

$$y(t) = y(0) \quad \text{for all } t \geq 0,$$

and

$$\lim_{t \to \infty} y(t) = y(0) = c^T x^0.$$

On the other hand,

$$\lim_{t \to \infty} y(t) = \lim_{t \to \infty} c^T x(t) = c^T (a\mathbf{1}) = a(c^T \mathbf{1}).$$

Hence, we obtain

$$a(c^T \mathbf{1}) = c^T x^0$$

and so

$$a = \frac{c^T x^0}{c^T \mathbf{1}}.$$

Now we have

$$\lim_{t \to \infty} x(t) = \left(\frac{c^T x^0}{c^T \mathbf{1}} \right) \mathbf{1}.$$

($\Longrightarrow$) To prove the contrapositive form, assume that the interaction graph $\mathcal{G}$ is not quasi-strongly connected. Then it follows from Theorem 2.1 that $\mathcal{G}_A$ has $k \geq 2$ closed strong components. This means that A has a zero eigenvalue of algebraic and geometric multiplicity $k \geq 2$ by Theorem 4.4. So we could find an initial state x^0 in $\ker(A)$ but not in Ω. This x^0 can always be found since the dimension of $\ker(A)$ is greater than 1. Further, for this initial state x^0, the solution will remain at x^0 for all t, i.e., $x(t) = x^0$. Hence, the system (5.8) does not achieve state agreement. ∎

Now we come to the case when $m > 1$. The linear coupled cell system (5.7) is written in matrix form as follows:

$$\dot{x}(t) = (A \otimes I_m)\, x(t). \tag{5.9}$$

Corollary 5.1. *The linear coupled cell system (5.9) achieves state agreement if and only if the interaction graph $\mathcal{G}$ is quasi-strongly connected. Moreover,*

$$\lim_{t \to \infty} x(t) = \mathbf{1} \otimes \left(\frac{(c^T \otimes I_2)x^0}{c^T \mathbf{1}} \right),$$

where c^T is a left-eigenvector of A corresponding to the zero eigenvalue.

Proof: Consider the $mn \times mn$ matrix partitioned as

$$I_{mn} = \begin{pmatrix} I_m & & \\ & \ddots & \\ & & I_m \end{pmatrix}.$$

Construct a permutation matrix P by selecting the rows of I_{mn} in the following order:

$$1, 1 + m, 1 + 2m, \ldots$$
$$2, 2 + m, 2 + 2m, \ldots$$
$$\vdots$$
$$m, 2m, 3m, \ldots$$

For example, for $m = 2$, $n = 3$

$$P = \begin{pmatrix} 1\,0\,0\,0\,0\,0 \\ 0\,0\,1\,0\,0\,0 \\ 0\,0\,0\,0\,1\,0 \\ 0\,1\,0\,0\,0\,0 \\ 0\,0\,0\,1\,0\,0 \\ 0\,0\,0\,0\,0\,1 \end{pmatrix}.$$

Observe that the matrix P performs the transformation

$$P(A \otimes I_m)P^T = I_m \otimes A = \begin{pmatrix} A & \cdots & 0 \\ \vdots & \ddots & \vdots \\ 0 & \cdots & A \end{pmatrix}.$$

Now applying the coordinate transformation $y = Px$ leads to the following new system

$$\dot{y}(t) = (I_m \otimes A)\, y(t).$$

Noticing the form of $I_m \otimes A$, we can directly apply Theorem 5.3 to obtain

$$\lim_{t \to \infty} y(t) = a \otimes \mathbf{1},$$

for some $a \in \mathbb{R}^m$. Thus,

$$\lim_{t \to \infty} x(t) = \lim_{t \to \infty} P^T y(t) = P^T(a \otimes \mathbf{1}) = \mathbf{1} \otimes a.$$

Hence, the linear coupled cell system (5.9) achieves state agreement if and only if the interaction graph $\mathcal{G}$ is quasi strongly connected.

Furthermore,

$$a = \frac{(c^T \otimes I_2)x^0}{c^T \mathbf{1}},$$

which can be easily obtained from Theorem 5.3.   ∎

5.4 Dynamic Topology: Symmetric Coupling

We now turn our focus to the harder problem: coupled cell systems with dynamic topology. In this section we shall address the case where all the system matrices are symmetric. In this case there is a common Lyapunov function.

Consider n cells and assume each cell's state is of dimension 1 without loss of generality since the higher dimensional case can also be treated similarly as in the preceding section. Thus consider a family of dynamic systems represented by the matrix form

$$\dot{x} = A_p x, \quad p \in \mathcal{P}$$

where $x \in \mathbb{R}^n$ is the aggregate state of n cells, $\mathcal{P}$ is a finite set, $A_p, p \in \mathcal{P}$, are generator matrices that are **symmetric**.

Given a piecewise constant switching signal $\sigma : \mathbb{R} \to \mathcal{P}$, we have the following switched linear coupled cell system

$$\dot{x}(t) = A_{\sigma(t)}x(t). \tag{5.10}$$

Let $\mathcal{G}(t)$ be the interaction graph for the above coupled cell system. Note that at any time t, the graph $\mathcal{G}(t)$ is bidirectional as the matrix $A_{\sigma(t)}$ is symmetric. Here again, let $x^0 = (x_1^0, \dots, x_n^0)$ be any initial state in $\mathbb{R}^n$.

We do not deal with fast chattering switching. So we assume all the switching signals are regular enough. Let $\mathcal{S}_{dwell}(\tau_D)$ denote the set of switching signals with *dwell time* $\tau_D > 0$, that is, two switching times differ by at least τ_D, a fixed constant. We shall show that if and only if a certain graphical condition holds, the switched linear coupled cell system above achieves state agreement. Indeed, the n cells' states will eventually converge to the centroid of their initial states.

Theorem 5.4. *Suppose that $\sigma(t) \in \mathcal{S}_{dwell}(\tau_D)$. The switched linear coupled cell system (5.10) achieves state agreement uniformly if and only if the interaction graph $\mathcal{G}(t)$ is uniformly quasi-strongly connected. Furthermore,*

$$\lim_{t \to \infty} x(t) = \left(\frac{x_1^0 + \cdots + x_n^0}{n} \right) \mathbf{1}.$$

The proof requires a lemma.

Lemma 5.1. *Suppose that $A \in \mathbb{R}^{n \times n}$ is a real symmetric matrix with eigenvalues λ_i satisfying*

$$\lambda_n \leq \lambda_{n-1} \leq \cdots \leq \lambda_k < \lambda_{k-1} = \cdots = \lambda_1 = 0.$$

Let $\mathcal{X}_0$ denote the zero eigenspace and let $\mathcal{X}_1$ denote the orthogonal complement of $\mathcal{X}_0$. Then for every $x \in \mathcal{X}_1$,

$$x^T A x \leq \lambda_k x^T x.$$

Proof: Let $c_1, c_2, \dots, c_n$ be normalized eigenvectors of A corresponding to the eigenvalues $\lambda_1, \lambda_2, \dots, \lambda_n$, respectively. Then

$$A = \lambda_1 c_1 c_1^T + \lambda_2 c_2 c_2^T + \cdots + \lambda_n c_n c_n^T = \lambda_k c_k c_k^T + \cdots + \lambda_n c_n c_n^T.$$

So for $x \in \mathcal{X}_1$,

$$x^T A x = \lambda_k x^T c_k c_k^T x + \cdots + \lambda_n x^T c_n c_n^T x$$
$$\leq \lambda_k x^T (c_k c_k^T + \cdots + c_n c_n^T) x$$
$$= \lambda_k x^T (c_1 c_1^T + c_2 c_2^T + \cdots + c_n c_n^T) x$$
$$= \lambda_k x^T x.$$

∎

Proof of Theorem 5.4: ($\Longleftarrow$) Let $\mathcal{X}_0 = \mathrm{span}\,\{\mathbf{1}\}$ and let $\mathcal{X}_1$ be the orthogonal complement of $\mathcal{X}_0$. Notice that for every $p \in \mathcal{P}$, A_p is a generator matrix and is symmetric. So it follows that

$$A_p \mathbf{1} = 0 \text{ and } \mathbf{1}^T A_p = 0, \text{ for all } p \in \mathcal{P}.$$

The trajectory of (5.10) looks like

$$x(t) = a(t)\mathbf{1} + w(t),$$

where $a(t) \in \mathbb{R}$ and $w(t) \in \mathcal{X}_1$. Pre-multiplying the equation above by $\mathbf{1}^T$, we get that
$$\mathbf{1}^T x(t) = na(t).$$
On the other hand, we have

$$\mathbf{1}^T \dot{x}(t) = \mathbf{1}^T A_{\sigma(t)} x(t) = 0.$$

So $a(t)$ is a constant real number, say $a(t) \equiv a$. Setting $t = t_0$, the starting time, we get $a = (x_1^0 + \cdots + x_n^0)/n$.

Now we come to show that $w(t) \to 0$ uniformly on t_0.

Since $w(t) = x(t) - a\mathbf{1}$ and $A_{\sigma(t)}\mathbf{1} = 0$,

$$\dot{w}(t) = \dot{x}(t) = A_{\sigma(t)} x(t) = A_{\sigma(t)} w(t). \tag{5.11}$$

We know that for any $w(t_0) \in \mathcal{X}_1$, the solution $w(t) \in \mathcal{X}_1$ for all $t \geq t_0$. In other words, $\mathcal{X}_1$ is a positively invariant set for the system (5.11).

Choose the positive definite function

$$V(w) = \frac{1}{2} w^T w.$$

Take the derivative of $V(w(t))$ along the solution of (5.11):

$$\dot{V}(w(t)) = w^T(t)A_{\sigma(t)}w(t).$$

By Theorem 4.2 we know $A_{\sigma(t)}$ is negative semi-definite. Hence, $V(w(t))$ is a non-increasing function with respect to t. Also it is lower-bounded by 0, so it has a limit as $t \to \infty$. Let

$$\lim_{t\to\infty} V(w(t)) = \varepsilon.$$

Now it remains to show that $\varepsilon = 0$. Suppose by contradiction that $\varepsilon > 0$. This implies

$$\|w(t)\|^2 \geq 2\varepsilon \quad \text{for all } t \geq t_0. \tag{5.12}$$

If the interaction graph $\mathcal{G}(t)$ is uniformly quasi-strongly connected, it follows that there exists a $T > 0$ such that for every $t \geq 0$ the union digraph $\mathcal{G}([t, t+T])$ is quasi-strongly connected.

Suppose $\sigma(t)$ switches at time instants

$$t_0, t_1, t_2, \ldots.$$

Now we generate a new subsequence $\{T_k\}$ from the sequence above as follows:

1. Set $T_0 = t_0$;
2. If $t_0 + T \in (t_{i-1}, t_i]$, set $T_1 = t_i$;
3. If $t_1 + T \in (t_{i-1}, t_i]$, set $T_2 = t_i$;
4. And so on.

By this construction, for every $k \geq 0$, we have $T_{k+1} - T_k \geq T$. This implies that the union digraph $\mathcal{G}([T_k, T_{k+1}])$ is quasi-strongly connected. Let the mode graphs during the interval $[T_k, T_{k+1}]$ be $\mathcal{G}_{k_1}, \ldots, \mathcal{G}_{k_\nu}$ and let the corresponding matrix be $A_{k_1}, \ldots, A_{k_\nu}$. Thus, the union digraph

$$\mathcal{G}_{k_1} \cup \cdots \cup \mathcal{G}_{k_\nu} = \mathcal{G}([T_k, T_{k+1}]).$$

Now we claim that for any time interval $[T_k, T_{k+1}]$, $w(t) \neq 0$ is not always in the null space of $A_{\sigma(t)}$. To see this, suppose by contradiction that $w(t) \neq 0$ is always in the null space of $A_{\sigma(t)}$ during $[T_k, T_{k+1}]$. That is,

$$A_{\sigma(t)}w(t) = 0.$$

Hence, $w(t) = w(T_k)$ for all $t \in [T_k, T_{k+1}]$ and so

$$A_{k_1} w(T_k) = 0, \quad \ldots, \quad A_{k_\nu} w(T_k) = 0.$$

This leads to

$$(A_{k_1} + \cdots + A_{k_\nu}) w(T_k) = 0. \tag{5.13}$$

Recall that each associated digraph $\mathcal{G}_{A_{k_j}}$ is the opposite digraph of $\mathcal{G}_{k_j}$. So the associated digraph $\mathcal{G}_{(A_{k_1}+\cdots+A_{k_\nu})}$ is just the opposite digraph of $\mathcal{G}_{k_1} \cup \cdots \cup \mathcal{G}_{k_\nu}$. Then we know $\mathcal{G}_{(A_{k_1}+\cdots+A_{k_\nu})}$ has one and only one closed strong component since the opposite digraph is quasi-strongly connected. Applying Theorem 4.4 we obtain the null space of $A_{k_1} + \cdots + A_{k_\nu}$ is

$$\text{span}\{\mathbf{1}\} = \mathcal{X}_0.$$

Combining with (5.13) we obtain $w(T_k) \in \mathcal{X}_0$, which contradicts $w(T_k) \in \mathcal{X}_1$. Therefore, there is $t_i \in [T_k, T_{k+1}]$ such that $w(t_i)$ is not in the null space of $A_{\sigma(t_i)}$. Recall that $A_{\sigma(t)} = A_{\sigma(t_i)}$ for all $t \in [t_i, t_{i+1}]$ since t_i is a switching time. Clearly, by the construction of the sequence $\{T_i\}$ we know t_{i+1} is also in $[T_k, T_{k+1}]$. Furthermore, the assumption $\sigma(t) \in \mathcal{S}_{dwell}(\tau_D)$ implies that

$$[t_i, t_i + \tau_D] \subset [t_i, t_{i+1}] \subset [T_k, T_{k+1}].$$

Here, we say $\sigma(t_i) = k_j$ for some $k_j \in \mathcal{P}$. Thus, applying Lemma 5.1 gives that, during $[t_i, t_i + \tau_D]$

$$\dot{V}(w(t)) = w^T(t) A_{\sigma(t)} w(t) = w^T(t) A_{k_j} w(t) \le \bar{\lambda}_{k_j} \|w(t)\|^2,$$

where $\bar{\lambda}_{k_j} \le 0$ is the largest nonzero eigenvalue of A_{k_j}. Considering (5.12), we obtain

$$\dot{V}(w(t)) \le 2\varepsilon \bar{\lambda}_{k_j}$$

and so

$$V\big(w(T_{k+1})\big) - V\big(w(T_k)\big) = \int_{T_k}^{T_{k+1}} \dot{V}(w(t))dt \le \int_{t_i}^{t_i+\tau_D} \dot{V}(w(t))dt \le 2\tau_D \varepsilon \bar{\lambda}_{k_j}.$$

Let $\bar{\lambda}$ be the maximum of the largest nonzero eigenvalues of $A_p, p \in \mathcal{P}$. Then

$$V(w(T_k)) \le V(w(T_0)) + 2k\tau_D \varepsilon \bar{\lambda}$$

and

$$\lim_{k \to \infty} V(w(T_k)) = -\infty,$$

which contradicts with $V(w(T_k))$ is nonnegative. Hence,

$$\lim_{t\to\infty} w(t) = 0 \quad \text{and} \quad \lim_{t\to\infty} x(t) = a\mathbf{1}.$$

Furthermore, notice that the convergence rate does not depend on initial time t_0 but depends only on $\bar{\lambda}$. Hence, the switched linear coupled cell system (5.10) achieves state agreement uniformly.

($\Longrightarrow$) The necessary proof is the same as the one for Theorem 5.6.   ■

5.5 Dynamic Topology: Asymmetric Coupling

In the previous section, we studied the switched linear coupled cell system where every system matrix is assumed to be symmetric. Now in this section we consider the general case without that assumption. A common Lyapunov function is not available any more for the general case. We have to find a new technique to prove the result though the statement of the theorem is still true.

Suppose we are given a family of dynamic systems represented by the matrix form

$$\dot{x} = A_p x, \quad p \in \mathcal{P}$$

where $x \in \mathbb{R}^n$ is the aggregate state of n cells and $\mathcal{P}$ is a finite set. Here the matrices A_p are not necessarily symmetric, but of course they are generator matrices.

For a piecewise constant switching signal $\sigma : \mathbb{R} \to \mathcal{P}$, the switched linear coupled cell system is given by

$$\dot{x}(t) = A_{\sigma(t)} x(t). \tag{5.14}$$

Let $\mathcal{G}(t)$ again be the interaction graph for the switched linear coupled cell system above and let x^0 be any initial state in $\mathbb{R}^n$.

Note in the stability result next that no assumption is needed about the interaction graph $\mathcal{G}(t)$.

Theorem 5.5. *Every agreement state is uniformly stable for the switched linear coupled cell system (5.14).*

Proof: Suppose the switching signal $\sigma(t)$ switches at

$$t_0, \ t_1, \ t_2, \ \dots.$$

Notice that for $t \in [t_i, t_{i+1}]$, $A_{\sigma(t)} = A_{\sigma(t_i)}$. So we let the transition matrix be

$$\Phi(t, t_i) = \exp\left(A_{\sigma(t_i)}(t - t_i)\right) \quad \text{for all } t \in [t_i, t_{i+1}].$$

Hence, the solution can be expressed by

$$x(t) = \Phi(t, t_i)\Phi(t_i, t_{i-1}) \cdots \Phi(t_1, t_0)x^0. \tag{5.15}$$

Since for each $p \in \mathcal{P}$, the matrix A_p is a generator matrix, by Theorem 4.1, all the transition matrices are stochastic and so is the product. Hence, it directly follows from (5.15) that, for any $t \geq t_0$ and for any $i = 1, \dots, n$, $x_i(t)$ is the convex combination of $x_1^0, \dots, x_n^0$ and therefore it is in the convex hull $\mathrm{co}\{x_1^0, \dots, x_n^0\}$.

In addition, notice that any equilibrium $\bar{x} \in \Omega$ is of the form $\bar{x} = a\mathbf{1}$. Let $\varepsilon > 0$ be arbitrary and choose $\delta = \varepsilon$. Thus for any t_0,

$$\|x^0 - \bar{x}\|_\infty \leq \delta \iff (\forall i) \ \|x_i^0 - a\|_\infty \leq \delta.$$

Clearly, for any x^0 satisfying $\|x^0 - \bar{x}\|_\infty \leq \delta$,

$$\mathrm{co}\{x_1^0, \dots, x_n^0\} \subset \{x \in \mathbb{R}^n : (\forall i) \ \|x_i^0 - a\|_\infty \leq \delta\}.$$

On the other hand, we know for every $t \geq t_0$

$$\mathrm{co}\{x_1(t), \dots, x_n(t)\} \subseteq \mathrm{co}\{x_1^0, \dots, x_n^0\}.$$

It follows now that

$$\|x(t) - \bar{x}\|_\infty \leq \varepsilon.$$

Hence, every agreement state $\bar{x} \in \Omega$ is uniformly stable for the switched linear coupled cell system (5.14). ∎

The graphical condition for achieving state agreement is the same as the one for the symmetric coupling case. Nevertheless, we cannot predict what common state the n cells will eventually converge to as in the symmetric coupling case. It depends not only on the initial state but also on the switching signal for the asymmetric coupling case.

Theorem 5.6. *Suppose that $\sigma(t) \in \mathcal{S}_{dwell}(\tau_D)$. The switched linear coupled cell system (5.14) achieves state agreement uniformly if and only if the interaction graph $\mathcal{G}(t)$ is uniformly quasi-strongly connected.*

Proof: ($\Longleftarrow$) Suppose the switching signal $\sigma(t)$ switches at

$$t_0, \ t_1, \ t_2, \ \ldots .$$

If there is only a finite number of switches, the final at time t_m, artificially define $t_{m+j} = t_m + jb$, $j = 1, 2, \ldots,$, where b is a finite positive value. Since $\sigma(t) \in \mathcal{S}_{dwell}(\tau_D)$, clearly,

$$t_{i+1} - t_i \geq \tau_D \quad \text{for all } i \geq 0.$$

In addition, we can always find a $\tau_m > \tau_D$ large enough that

$$t_{i+1} - t_i \leq \tau_m \quad \text{for all } i \geq 0,$$

since otherwise if there is no such τ_m for some interval $[t_i, t_{i+1}]$ we can artificially partition it.

If the interaction graph $\mathcal{G}(t)$ is uniformly quasi-strongly connected, by definition there is a positive T such that for all $t \geq 0$ the union digraph $\mathcal{G}([t, t+T])$ is quasi-strongly connected. Now we generate a subsequence $\{t_{m_j}\}$ of the sequence $\{t_i\}$ as follows:

(1) Set $m_0 = 0$.

(2) If $t_{m_0} + T \in (t_{i-1}, t_i]$, set $m_1 = i$.

(3) If $t_{m_1} + T \in (t_{i-1}, t_i]$, set $m_2 = i$.

(4) And so on.

Notice that for $t \in [t_i, t_{i+1}]$, $A_{\sigma(t)} = A_{\sigma(t_i)}$. So we let the transition matrix be

$$\Phi(t, t_i) = \exp\left(A_{\sigma(t_i)}(t - t_i)\right) \quad \text{for all } t \in [t_i, t_{i+1}].$$

Then it is clear that

$$x(t_{m_1}) = \Phi(t_{m_1}, t_{m_1-1})\Phi(t_{m_1-1}, t_{m_1-2}) \cdots \Phi(t_1, t_0)x^0$$
$$=: \Psi_1 x^0,$$

$$x(t_{m_2}) = \Phi(t_{m_2}, t_{m_2-1})\Phi(t_{m_2-1}, t_{m_2-2}) \cdots \Phi(t_{m_1+1}, t_{m_1})x(t_{m_1})$$
$$=: \Psi_2 x(t_{m_1}) = \Psi_2 \Psi_1 x^0,$$

$$\vdots$$

$$x(t_{m_j}) = \Phi(t_{m_j}, t_{m_j-1})\Phi(t_{m_j-1}, t_{m_j-2}) \cdots \Phi(t_{m_{j-1}+1}, t_{m_{j-1}})x(t_{m_{j-1}})$$
$$=: \Psi_j x(t_{m_{j-1}}) = \Psi_j \Psi_{j-1} \cdots \Psi_1 x^0,$$

$$\vdots$$

Firstly, we will show $\lim_{j \to \infty} x(t_{m_j}) = a\mathbf{1}$ for some $a \in \mathbb{R}$. It suffices to show that

$$\lim_{j \to \infty} \Psi_j \Psi_{j-1} \cdots \Psi_1 = \mathbf{1}c^T,$$

where c^T is a row vector. (Then $a = c^T x^0$.)

Recall that for each $p \in \mathcal{P}$, A_p is a generator matrix. So it can be written as

$$A_p = -s_p I + E_p,$$

where s_p is a positive scalar large enough that E_p is nonnegative. Hence,

$$\Phi(t_{i+1}, t_i) = \exp\left(A_{\sigma(t_i)}(t_{i+1} - t_i)\right) = \exp\left\{\left(-s_{\sigma(t_i)}I + E_{\sigma(t_i)}\right)(t_{i+1} - t_i)\right\}$$
$$= e^{-s_{\sigma(t_i)}(t_{i+1}-t_i)}\left\{I + E_{\sigma(t_i)}(t_{i+1} - t_i) + \frac{E_{\sigma(t_i)}^2 (t_{i+1}-t_i)^2}{2!} + \cdots\right\},$$

and

$$\Psi_j = \Phi(t_{m_j}, t_{m_j-1})\Phi(t_{m_j-1}, t_{m_j-2}) \cdots \Phi(t_{m_{j-1}+1}, t_{m_{j-1}}) = \alpha_j(I + \Theta_j + \Gamma_j),$$

where

$$\alpha_j = e^{-s_{\sigma(t_{m_j-1})}(t_{m_j}-t_{m_j-1})} \cdots e^{-s_{\sigma(t_{m_{j-1}})}(t_{m_{j-1}+1}-t_{m_{j-1}})},$$
$$\Theta_j = E_{\sigma(t_{m_j-1})}(t_{m_j} - t_{m_j-1}) + \cdots E_{\sigma(t_{m_{j-1}})}(t_{m_{j-1}+1} - t_{m_{j-1}}),$$

and Γ_j is a nonnegative matrix representing the sum of all the other terms.

First of all, by Theorem 4.1, we know, for every i, $\Phi(t_{i+1}-t_i)$ is stochastic and so is Ψ_j for every j. Secondly, for each $p \in \mathcal{P}$, recalling that $\mathcal{G}_{A_p}$ is an opposite digraph of the mode graph $\mathcal{G}_p$, it is clear that

$$\mathcal{G}(\Theta_j) = \mathcal{G}\left(E_{\sigma(t_{m_j}-1)}\right) \cup \cdots \cup \mathcal{G}\left(E_{\sigma(t_{m_j-1})}\right)$$

has a globally reachable node since $\mathcal{G}\left([t_{m_{j-1}}, t_{m_j}]\right)$ is quasi-strongly connected by construction. Furthermore, $\mathcal{G}(I)$ is a digraph with loop on each node. So

$$\mathcal{G}(\Psi_j) = \mathcal{G}(I) \cup \mathcal{G}(\Theta_j) \cup \mathcal{G}(\Gamma_j)$$

has a globally reachable node which is aperiodic. Hence it follows from Theorem 3.7 that Ψ_j is SIA for every j.

Let

$$\Xi = \{\Psi_1, \Psi_2, \dots\}.$$

Thus, Ξ is a set of SIA matrices which are of order n. Let l be the number of different associated digraphs of all SIA matrices of the same order n. Similar to the argument above, we can show that every product of some Ψ_j given as the following

$$\Psi_{k_1}\Psi_{k_2}\cdots\Psi_{k_i} = \alpha_{k_1}(I + \Theta_{k_1} + \Gamma_{k_1})\alpha_{k_2}(I + \Theta_{k_2} + \Gamma_{k_2})\cdots\alpha_{k_i}(I + \Theta_{k_i} + \Gamma_{k_i})$$
$$= \alpha_{k_1}\alpha_{k_2}\cdots\alpha_{k_i}(I + \Theta_{k_1} + \Theta_{k_2} + \cdots + \Theta_{k_i} + \cdots)$$

is still SIA. Then it follows from Lemma 3.4 that all products in Ξ of length $l+1$ are scrambling matrices. Notice that $\mathcal{P}$ is a finite set. So the positive entries of all the matrices $A_p, p \in \mathcal{P}$ have a uniform lower-bound. This leads to the fact that all positive entries of Ψ_j have a uniform lower-bound, which again implies that the positive entries of any products in Ξ of length $l+1$ have a uniform lower-bound. Combing the fact above and the fact that any product in Ξ of length $l+1$ is a scrambling matrix, we obtain that there is a d ($0 \le d < 1$) such that for any product in Ξ of length $l+1$,

$$\lambda\left(\Psi_{k_1}\Psi_{k_2}\cdots\Psi_{k_{m+1}}\right) \le d,$$

where $\lambda(\cdot)$ is a scalar function defined in (3.15). Furthermore, notice that d is independent of the choice of t_0 but only depends on the lower-bound of the nonnegative entries, which are related to τ_D, τ_m, and the matrices $A_p, p \in \mathcal{P}$. Thus applying Theorem 3.8 (Wolfowitz theorem) gives

$$\lim_{j\to\infty} \Psi_j \Psi_{j-1} \cdots \Psi_1 = \mathbf{1}c^T$$

uniformly on t_0, which then immediately leads to

$$\lim_{j\to\infty} x(t_{m_j}) = \mathbf{1}c^T x^0 = a\mathbf{1}$$

uniformly on t_0.

Now we come to look at the times between the instants $\{t_{m_0}, t_{m_1}, t_{m_2}, \dots\}$. For $t \in [t_{m_j}, t_{m_{j+1}}]$,

$$x(t) = \Phi(t, t_{m_j+k})\Phi(t_{m_j+k}, t_{m_j+k-1})\cdots\Phi(t_{m_j+1}, t_{m_j})x(t_{m_j}) =: \widetilde{\Phi}x(t_{m_j}),$$

where $\widetilde{\Phi}$ is also stochastic for the same reason. This means that each cell's state $x_i(t)$, $i = 1, 2, \dots, n$ is a convex combination of $x_1(t_{m_j}), \dots, x_m(t_{m_j})$. Hence,

$$\lim_{t\to\infty} x(t) = a\mathbf{1}$$

uniformly on t_0. In conclusion, the switched linear couple cell system (5.14) achieves state agreement uniformly.

($\Longrightarrow$) To prove the contrapositive form, assume that $\mathcal{G}(t)$ is not uniformly quasi-strongly connected. That is, for all $T > 0$ there exists $t^* \geq 0$ such that the digraph $\mathcal{G}([t^*, t^* + T])$ is not quasi-strongly connected. During the interval $[t^*, t^* + T]$, let $\sigma(t)$ takes values $p_1, p_2, \dots, p_k$ in $\mathcal{P}$. This implies

$$\mathcal{G}_{p_1} \cup \mathcal{G}_{p_2} \cup \cdots \cup \mathcal{G}_{p_k}$$

is not quasi-strongly connected. Recall that the mode graph $\mathcal{G}_p$ is just the opposite digraph of $\mathcal{G}_{A_p}$. So the associated digraph

$$\mathcal{G}_{\left(A_{p_1}+A_{p_2}+\cdots+A_{p_k}\right)}$$

has at least two closed strong components by Theorem 2.1. This implies that

$$A_{p_1}, A_{p_2}, \dots, A_{p_k}$$

share a common null space of dimension at least 2. Thus we can find an initial state x^0 in this common null space but not in Ω. Let $c = \|x^0\|$ and let $\varepsilon = \|x^0\|_\Omega$, where $\|\cdot\|_\Omega$ represents the distance from a point to the set Ω. Clearly,

$$c > 0, \ \varepsilon > 0.$$

Notice that for all T there exists a $t_0 = t^*$ such that if $x(t_0) = x^0$ then $x(t) = x^0$ for all $t \in [t_0, t_0 + T]$. So

$$(\exists t = t_0 + T) \ \|x(t)\|_\Omega = \varepsilon,$$

which means the switched linear coupled cell system (5.14) does not achieve state agreement uniformly.   ∎

As an example, a simulation of linear coupled cell system with dynamic topology is given in Fig. 5.4. Each cell has its state in $\mathbb{R}^2$. The interaction graph is drawn in the picture using dash lines with arrows at different time instants. The system achieves state agreement as shown.

5.6 Switched Positive Systems

The result developed in the proceeding sections can be used to check whether a class of switched positive systems is asymptotically stable with respect to the origin.

Consider a family of system matrices $\{A_p : p \in \mathcal{P}\}$ where all the matrices are Metzler matrices with row sums less than or equal to zero. Given a switching signal $\sigma : \mathbb{R} \to \mathcal{P}$, it gives rise to a switched system

$$\dot{x}(t) = A_{\sigma(t)}x(t), \tag{5.16}$$

where $x \in \mathbb{R}^n$ is its state. This system is called a positive system and is studied quite often without switching and has wide applications in social sciences and some other areas [3, 39, 67, 86].

Treat (5.16) as a coupled cell system of n agents with each agent's state x_i, $i = 1, \ldots, n$. Now artificially construct a virtual agent (with state x_0) at the origin and keep it stationary. In other words,

$$\dot{x}_0 = 0, \qquad x_0^0 = 0.$$

Then for any $p \in \mathcal{P}$, we can write a new system consisting of $n + 1$ agents as follows

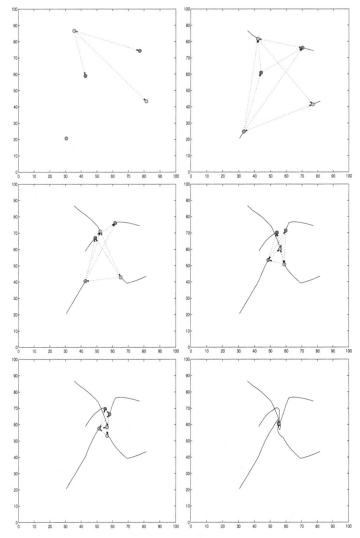

Fig. 5.4. Simulation of linear coupled system with dynamic topology.

$$\begin{pmatrix} \dot{x}_0 \\ \dot{x} \end{pmatrix} = \left(\begin{array}{c|c} 0 & 0 \\ \hline -\sum_{j=1}^{n} a_{1j}(p) & \\ \vdots & A_p \\ -\sum_{j=1}^{n} a_{nj}(p) & \end{array} \right) \begin{pmatrix} x_0 \\ x \end{pmatrix}, \tag{5.17}$$

where $a_{ij}(p)$ denotes the (i,j)th entry of A_p. Let $\bar{A}_p$ denote the system matrix of the above new system. Notice that $-\sum_{j=1}^{n} a_{ij}(p)$ is nonnegative. So $\bar{A}_p$ is a generator matrix for any $p \in \mathcal{P}$. Furthermore, notice that the solution trajectory $x(t)$ of (5.16) is just the partial solution $x(t)$ of the following switched system

$$\begin{pmatrix} \dot{x}_0(t) \\ \dot{x}(t) \end{pmatrix} = \bar{A}_{\sigma(t)} \begin{pmatrix} x_0(t) \\ x(t) \end{pmatrix} \tag{5.18}$$

for any initial condition x^0 and $x_0^0 = 0$. Hence, the switched system (5.16) is globally uniformly asymptotically stable with respect to the origin if and only if the new switched system (5.18) achieves state agreement uniformly.

Define the interaction graph $\mathcal{G}(t)$ for these $n+1$ agents in terms of $\bar{A}_{\sigma(t)}$. Thus we can find out whether the switched system (5.16) is globally uniformly asymptotically stable with respect to the origin by checking whether $\mathcal{G}(t)$ is uniformly quasi-strongly connected.

As an example, take $\mathcal{P} = \{1, 2\}$ and consider the two matrices

$$A_1 = \begin{pmatrix} -3 & 2 \\ 0 & 0 \end{pmatrix} \quad \text{and} \quad A_2 = \begin{pmatrix} 0 & 0 \\ 1 & -2 \end{pmatrix}.$$

Further, give a periodic piecewise constant switching signal shown at the bottom of Fig. 5.6. Thus we have a switched system of the form (5.16).

Now whether this switched system is globally uniformly asymptotically stable or not can be determined from its structure. Treat the system as a coupled cell system composed of two agents labelled 1 and 2 which correspond to the state components x_1 and x_2. Add a virtual agent labelled 0 and get an augmented system of the form (5.17) for $p \in \mathcal{P}$. Correspondingly, at each mode $p = 1, 2$, the mode graphs $\mathcal{G}_1$ and $\mathcal{G}_2$ can be easily obtained and are depicted in Fig. 5.5. Notice that neither $\mathcal{G}_1$ nor $\mathcal{G}_2$ is quasi-strongly connected, but the interaction graph $\mathcal{G}(t)$ is uniformly quasi-strongly connected for the switching signal given at the bottom of Fig. 5.6. Hence, by our result, the three agents globally uniformly converge to a same point. In addition, note that the virtual agent is stationary at the origin. This means that the original switched system is globally uniformly

Fig. 5.5. Mode graphs $\mathcal{G}_1$ and $\mathcal{G}_2$.

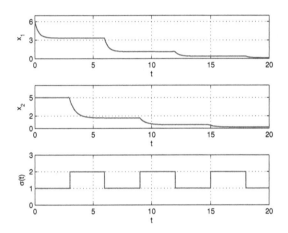

Fig. 5.6. Time response and switching signal.

asymptotically stable with respect to the origin. The time responses and phase portrait are depicted in Fig. 5.6 and Fig. 5.7, respectively.

5.7 Notes and Discussion

The study of interaction among cells (agents) plays an important role in understanding the collective behavior—agreement at the macro level. The first effort in this direction is reported in [33]. Consider a group of individuals who must act together as a team or committee, and suppose that each individual in the group has his own subjective probability distribution for the unknown value of some parameter. A linear coupled system model in discrete time is presented [33] which describes how the group might reach agreement on a common subjective probability distribution for the parameter by pooling individuals' opinions. This

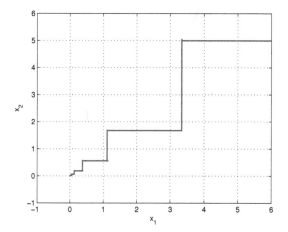

Fig. 5.7. Asymptotically stable trajectory.

is also related to the Delphi method which was originally developed at the RAND Corporation by Olaf Helmer and Norman Dalkey.

Agreement problem (or consensus problem) also appears in the field of distributed computing [87]. Distributed computation over networks has a tradition in systems and control theory starting with the work of [24] and [136] on asynchronous asymptotic agreement problem for distributed decision-making systems.

Recently, in many applications involving multi-agent systems, the interaction topology between agents may change dynamically. For instance, links may be dropped and formed due to communication disturbances and/or subject to sensor range limitations. Vicsek et al. [137] provides an example. In that work, they propose a simple but compelling discrete-time model of n autonomous agents (i.e., particles) all moving in the plane with the same speed but with different headings. Each agent updates its heading using a nearest neighbor rule based on the average of the headings of itself and its neighbors. Each agent's neighbors are defined to be those agents in the disk of a pre-specified radius centered at its current location. Thus, the nearest neighbor rule leads to a dynamically changing topology among them. In their paper, the authors provide a variety of interesting simulation results which demonstrate that this rule can cause all agents to eventually move in the same direction—an agreement of the heading is achieved. Then [58] attempts to provide a formal analysis for this observed behavior. The authors study the Vicsek discrete-time model in the framework of switched system (that is, the system matrix switches among a finite family of stochastic matrices), where the interaction structure of each mode is modeled by an undirected graph. To analyze such models, the authors adopt a Wolfowitz's theorem (a convergence property of infinite products of certain type of stochastic matrices) from ergodic theory to prove that all n agents' headings converge to a common steady state provided that the union of undirected graphs is connected with sufficient frequency. Later on, an extension is made in [94] to allow for time-dependent, bidirectional and unidirectional interactions.

In contrast to the discrete time setup, much research work in continuous time emerges for linear coupled systems with fixed and/or dynamic topologies in recent years. In [102, 103, 118], the common Lyapunov function technique is used to show that strongly connected and balanced digraphs play a key role in addressing the average-consensus problem for linear coupled systems. In [13, 107], the problem of information consensus among multiple agents under fixed and dynamic communication links is studied. Two other works on this problem are [53], which deals with the agreement problem over random information networks, and [95], which addresses the deterministic time-varying case where a sufficient condition for convergence is presented and robustness with respect to an arbitrary delay is also taken into account. In addition, [123] and [61] study the agreement problem of a group of agents, agreement of speed, position, etc., when the data is quantized in amplitude. More discussion on linear consensus problem can be found in several survey papers (e.g., [101, 108]).

In this chapter, a general abstract model for linear coupled cell systems is established in continuous time, which encompasses the models studied in [13, 103, 109, 118]. A complete and systematic analysis is developed for the stability properties of agreement state. With differences regarding the coupling structure, we analyze four cases: i) cyclic coupling structure, ii) arbitrary coupling structure but fixed, iii) symmetric coupling and dynamic topology, iv) asymmetric coupling and dynamic topology. Necessary and sufficient conditions to assure state agreement are obtained in each case.

In discrete-time, linear coupled cell systems can be summarized as follows:

$$x_i(k+1) = \sum_{j \in \mathcal{N}_i(k)} a_{ij} x_j(k), \ 0 \le a_{ij} \le 1, \quad i = 1, \ldots, n$$

or in matrix form

$$x(k+1) = \left(A_{\sigma(k)} \otimes I_m\right) x(k).$$

The family of matrices $\{A_\sigma\}$ are assumed to be row stochastic with positive diagonal entries. Then the state agreement occurs if and only if the interaction graph is uniformly quasi-strongly connected.

6

Nonlinear Coupled Cell Systems

This chapter studies nonlinear coupled cell systems with fixed and dynamic topologies. The vector fields of the individual systems are assumed to satisfy certain hypotheses. Again, the equilibrium set contains all states with identical state components. This class of systems generalizes that of linear coupled cell systems and is abundant in biology, physics, engineering, ecology, and social science: e.g., a biochemical reaction network [49], coupled oscillators [59, 127], arrays of chaotic systems [138, 144–146], and a swarm of organisms [46, 47]. We model such systems by coupled nonlinear differential equations in state form. Similar to the linear case, central to the stability issue of such systems is the graph describing the interaction structure—that is, who is coupled to whom?

6.1 State Model and Interaction Graph

In this section we introduce a general nonlinear state model to describe nonlinear coupled cell systems with either fixed topology or dynamic topology. Examples and applications are discussed in later sections.

Suppose that we are given a family of systems represented by nonlinear ordinary differential equations of the form

$$\begin{aligned}
\dot{x}_1 &= f_p^1(x_1, \dots, x_n) \\
&\vdots \\
\dot{x}_n &= f_p^n(x_1, \dots, x_n)
\end{aligned} \tag{6.1}$$

where $x_i \in \mathcal{X}$ is the state of *cell* or *agent* i, $i = 1, \dots, n$, and where the index p lives in a finite set $\mathcal{P}$. Notice that all the cells share a common $\mathcal{X}$. (Here $\mathcal{X}$ can be an Euclidean space $\mathbb{R}^m$ or just a set in $\mathbb{R}^m$.)

Introducing the *aggregate state* $x = (x_1, \ldots, x_n) \in \mathcal{X}^n$, we have the concise form

$$\dot{x} = f_p(x), \quad p \in \mathcal{P} \tag{6.2}$$

where $f_p : \mathcal{X}^n \to \mathcal{X}^n$, $p \in \mathcal{P}$, is a family of sufficiently regular (continuous or locally Lipschitz) vector fields, parameterized by the index set $\mathcal{P}$. Each individual component model $\dot{x} = f_p(x)$ for $p \in \mathcal{P}$ is called a *mode* of the family (6.2).

Similar to the linear case, for a piecewise constant switching signal $\sigma : \mathbb{R} \to \mathcal{P}$, which specifies, at each time instant t, the index $\sigma(t) \in \mathcal{P}$ of the active mode, we thus have the following *switched nonlinear coupled cell system*

$$\dot{x}(t) = f_{\sigma(t)}\left(x(t)\right). \tag{6.3}$$

We assume that the state of the system above does not jump at switching times, i.e., the solution $x(\cdot)$ is everywhere continuous.

We now turn to define the *interaction graph* for the nonlinear coupled cell system above.

Definition 6.1. *The* interaction graph $\mathcal{G}(t) = (\mathcal{V}, \mathcal{E}(t))$ *consists of*

- *a finite set $\mathcal{V}$ of n nodes, each node i modeling cell i;*
- *a (time-varying) arc set $\mathcal{E}(t)$ representing the links between cells at time t. An arc from node j to node i at t indicates that cell j is a neighbor of cell i in the sense that $f^i_{\sigma(t)}$ depends on x_j, i.e., there exist $x^1_j, x^2_j \in \mathcal{X}$ such that*

$$f^i_{\sigma(t)}(x_1, \ldots, x^1_j, \ldots, x_n) \neq f^i_{\sigma(t)}(x_1, \ldots, x^2_j, \ldots, x_n).$$

The set of neighbors of cell i at t is denoted by $\mathcal{N}_i(t)$.

Like the linear case, the interaction graph $\mathcal{G}(t)$ captures the evolution of the coupling structure among the n cells according to the switching signal $\sigma(t)$. At each mode $p \in \mathcal{P}$, we define the *mode graph* $\mathcal{G}_p$ as

$$\mathcal{G}_p := \mathcal{G}(t) \text{ when } \sigma(t) \equiv p.$$

The next example combines several of the concepts presented thus far.

Suppose we are given a family of nonlinear coupled systems of three cells and two modes ($\mathcal{P} = \{1, 2\}$):

$$p = 1 : \begin{cases} \dot{x}_1 = f_1^1(x_1, x_2, x_3) \\ \dot{x}_2 = f_1^2(x_2, x_3) \\ \dot{x}_3 = f_1^3(x_1, x_3) \end{cases}, \quad p = 2 : \begin{cases} \dot{x}_1 = f_2^1(x_1, x_2) \\ \dot{x}_2 = f_2^2(x_1, x_2) \\ \dot{x}_3 = f_2^3(x_3) \end{cases}.$$

Further, suppose we are given a switching signal $\sigma : \mathbb{R} \to \mathcal{P}$ depicted in Fig. 6.1.

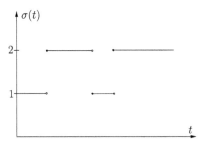

Fig. 6.1. Switching signal.

Thus a switched nonlinear coupled cell system with switching signal $\sigma(t)$ can be described by the equation

$$\dot{x}(t) = f_{\sigma(t)}\big(x(t)\big)$$

where $x = (x_1, x_2, x_3)$ is the aggregate state and

$$f_{\sigma(t)} = \big(f_{\sigma(t)}^1, f_{\sigma(t)}^2, f_{\sigma(t)}^3\big)$$

is the vector field.

The mode graphs $\mathcal{G}_1$ and $\mathcal{G}_2$ corresponding to the modes $p = 1$ and $p = 2$ are shown in Fig. 6.2. The nodes 1, 2, 3 represent the cells 1, 2, 3 and the arcs in the digraphs represent the coupling links in terms of Definition 6.1.

The interaction graph $\mathcal{G}(t)$ is a digraph of three nodes where the arc set changes over time depending on $\sigma(t)$.

Having introduced the switched nonlinear coupled cell system model and the interaction graph capturing the coupling structure of the system, we are now

Fig. 6.2. Mode graphs.

ready to give some formal definitions related to the problems that are going to be addressed in this chapter.

Assume $\Omega = \{x \in \mathcal{X}^n : x_1 = \cdots = x_n\}$ is an equilibria set of the switched nonlinear coupled cell system (6.3). Later, assumptions will be made for the existence of Ω. We call $\bar{x} \in \Omega$ an *agreement state*. Let $x^0 = (x_1^0, \ldots, x_n^0)$ be any initial state in $\mathcal{X}^n$ and let $x(t)$ be solutions of the coupled cell system (6.3) starting at (t_0, x^0) (i.e., $x(t_0) = x_0$).

Definition 6.2. *The nonlinear coupled cell system (6.3) achieves*

- state agreement in $\mathcal{X}$ *if* $\forall \varepsilon > 0, \forall t_0 \geq 0, \exists T > 0$ *such that*

$$(\forall i) \ x_i^0 \in \mathcal{X} \implies (\exists \varsigma \in \mathcal{X})(\forall t \geq t_0 + T)(\forall i) \ \|x_i(t) - \varsigma\| \leq \varepsilon;$$

- state agreement in $\mathcal{X}$ *uniformly if* $\forall \varepsilon > 0, \forall c > 0, \exists T > 0$ *such that* $\forall t_0 \geq 0$

$$(\forall i) \ x_i^0 \in \{y \in \mathcal{X} : \|y\| \leq c\} \implies (\exists \varsigma \in \mathcal{X})(\forall t \geq t_0 + T)(\forall i) \ \|x_i(t) - \varsigma\| \leq \varepsilon.$$

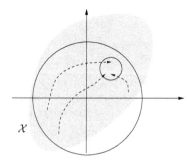

Fig. 6.3. Achieve state agreement in $\mathcal{X}$ uniformly.

The definition is illustrated in Fig. 6.3. Uniformity requires a uniform convergence rate with respect to starting time t_0 and initial state x^0 in a bounded set. There is no distinction between achieving state agreement and achieving state agreement uniformly for autonomous systems. When $\mathcal{X} = \mathbb{R}^m$, we say the system globally achieves state agreement (uniformly).

In the particular case when $\sigma(t)$ is a constant signal, that is, $\sigma(t) \equiv p$, the switched nonlinear coupled cell system turns out be time-invariant and the interaction graph $\mathcal{G}(t)$ is just a fixed interaction graph $\mathcal{G}_p$. We will drop the subscript p for simplicity. Therefore, with fixed topology, the nonlinear coupled cell system of interest is given by

$$\dot{x}_1 = f^1(x_1, \ldots, x_n)$$

$$\vdots \tag{6.4}$$

$$\dot{x}_n = f^n(x_1, \ldots, x_n)$$

or

$$\dot{x} = f(x)$$

in vector form. Correspondingly, the interaction graph is denoted by $\mathcal{G}$. Although it is a special case of (6.3), we study it in a different way for better understanding.

6.2 Tangent Cones

In this section we assemble some basic concepts, notations, and some properties regarding convex sets and tangent cones, which are used in the remainder of this chapter. The reader may refer to [8, 9, 113, 125] for details.

Let $\mathcal{S} \subset \mathbb{R}^m$. The intersection of all convex sets containing $\mathcal{S}$ is the *convex hull* of $\mathcal{S}$, denoted co($\mathcal{S}$). The convex hull of a finite set of points $x_1, \ldots, x_n \in \mathbb{R}^m$ is a *polytope*, denoted co$\{x_1, \ldots, x_n\}$.

We denote the interior of $\mathcal{S}$ by int($\mathcal{S}$) and the boundary of $\mathcal{S}$ by $\partial(\mathcal{S})$, respectively. If $\mathcal{S}$ contains the origin, the smallest subspace containing $\mathcal{S}$ is the *carrier subspace*, denoted lin($\mathcal{S}$). The *relative interior* of $\mathcal{S}$, denoted ri($\mathcal{S}$), is the interior of $\mathcal{S}$ when it is regarded as a subset of lin($\mathcal{S}$) and the relative topology is used. Likewise for the *relative boundary*, denoted rb($\mathcal{S}$). If $\mathcal{S}$ does not contain the origin, it must be translated by an arbitrary vector: Let v be any point in $\mathcal{S}$ and let lin($\mathcal{S}$) denote the smallest subspace containing $\mathcal{S} - v$. Then ri($\mathcal{S}$) is the interior of $\mathcal{S}$ when it is regarded as a subset of the affine subspace $v + \text{lin}(\mathcal{S})$.

In the trivial case, when $\mathcal{S}$ is just a point, the relative interior $\mathrm{ri}(\mathcal{S})$ is itself. In precise terms,

$$\mathrm{ri}(\mathcal{S}) = \left\{ y \in \mathcal{S} : \exists \, \epsilon > 0, \ y + \left(\epsilon \mathcal{B}_m \cap \mathrm{lin}(\mathcal{S}) \right) \subset \mathcal{S} \right\},$$

where $\mathcal{B}_m$ is the unit ball in $\mathbb{R}^m$; see Fig. 6.4. Similarly for $\mathrm{rb}(\mathcal{S})$.

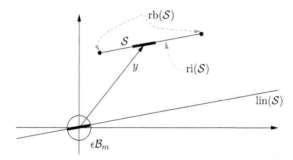

Fig. 6.4. Set $\mathcal{S}$, $\mathrm{lin}(\mathcal{S})$, $\mathrm{ri}(\mathcal{S})$, and $\mathrm{rb}(\mathcal{S})$.

Fix any norm $\|\cdot\|$ in $\mathbb{R}^m$. For each nonempty subset $\mathcal{S}$ of $\mathbb{R}^m$ and each $y \in \mathbb{R}^m$, we denote the distance of y from $\mathcal{S}$ by

$$\|y\|_{\mathcal{S}} := \inf_{z \in \mathcal{S}} \|z - y\|.$$

A nonempty set $\mathcal{K} \subset \mathbb{R}^m$ is called a *cone* if $\lambda y \in \mathcal{K}$ when $y \in \mathcal{K}$ and $\lambda > 0$. Let $\mathcal{S} \subset \mathbb{R}^m$ be a closed convex set and $y \in \mathcal{S}$. The *tangent cone* (often referred to as *contingent cone*, Bouligand, 1932, [25]) to $\mathcal{S}$ at y is the set

$$T(y, \mathcal{S}) = \left\{ z \in \mathbb{R}^m : \liminf_{\lambda \to 0} \frac{\|y + \lambda z\|_{\mathcal{S}}}{\lambda} = 0 \right\}$$

and the *normal cone* to $\mathcal{S}$ at y is

$$\mathcal{N}(y, \mathcal{S}) = \{ z^* : \langle z, z^* \rangle \leq 0, \ \forall z \in T(y, \mathcal{S}) \}.$$

Note that if $y \in \mathrm{int}(\mathcal{S})$, then $T(y, \mathcal{S}) = \mathbb{R}^m$. Thus the set $T(y, \mathcal{S})$ is non-trivial only on $\partial \mathcal{S}$. In particular, if $\mathcal{S}$ contains only one point, y, then $T(y, \mathcal{S}) = \{0\}$. In geometric terms (see Fig. 6.5), the tangent cone for $y \in \partial \mathcal{S}$ is a cone having

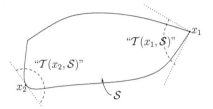

Fig. 6.5. Translated tangent cones $\mathcal{T}(x_1, \mathcal{S})$ and $\mathcal{T}(x_2, \mathcal{S})$.

center at the origin that contains all vectors whose directions point from y 'inside' (or they are 'tangent to') the set $\mathcal{S}$. If such a boundary is smooth at the point y, then the $\mathcal{T}(y, \mathcal{S})$ is just the tangent halfspace shifted to the origin. For example, in Fig. 6.5, the boundary at x_2 is smooth, whereas the boundary at x_1 is not smooth, so the $\mathcal{T}(x_2, \mathcal{S})$ is just the tangent halfspace shifted to the origin but the $\mathcal{T}(x_1, \mathcal{S})$ is not.

Before concluding this section, we summarize in the following lemma some properties of tangent cones to convex sets. A graphical interpretation of the lemma is given in Fig. 6.6. It is stated that if a convex set $\mathcal{S}_2$ contains another

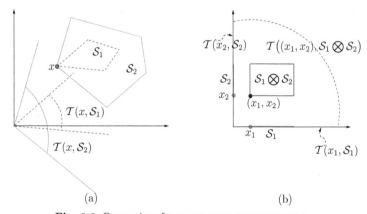

Fig. 6.6. Properties of tangent cones to convex sets.

convex set $\mathcal{S}_1$, then the tangent cone at any point x to the set $\mathcal{S}_2$ also contains the tangent cone at x to the set $\mathcal{S}_1$ (see Fig. 6.6(a)). In addition, if a convex set is a Cartesian product of n convex sets $\mathcal{S}_1, \ldots \mathcal{S}_n$, then the tangent cone at any point $x = (x_1, \ldots, x_n)$ to the set is just the Cartesian product of these tangent cones $\mathcal{T}(x_i, \mathcal{S}_i), i = 1, \ldots, n$. As an example, in Fig. 6.6(b), a convex set (square) in the plane is the Cartesian product of two intervals $\mathcal{S}_1$ and $\mathcal{S}_2$. Notice that the tangent cone at x_1 to the interval set $\mathcal{S}_1$ is just the positive horizontal axis and the tangent cone at x_2 to the interval set $\mathcal{S}_2$ is the positive vertical axis. Their Cartesian product is the first quadrant, which is exactly the tangent cone at the point (x_1, x_2) to the square set.

Lemma 6.1 ([8], page 164). *Let $\mathcal{S}_i, i = 1, \ldots, n$ be closed convex sets in $\mathbb{R}^m$. The following properties hold:*

1. *If $y \in \mathcal{S}_1 \subset \mathcal{S}_2$, then $\mathcal{T}(y, \mathcal{S}_1) \subset \mathcal{T}(y, \mathcal{S}_2)$ and $\mathcal{N}(y, \mathcal{S}_2) \subset \mathcal{N}(y, \mathcal{S}_1)$;*
2. *If $x_i \in \mathcal{S}_i \ (i = 1, \ldots, n)$, then*

$$\mathcal{T}\left((x_1, \ldots, x_n), \bigotimes_{i=1}^{n} \mathcal{S}_i \right) = \bigotimes_{i=1}^{n} \mathcal{T}(x_i, \mathcal{S}_i),$$

$$\mathcal{N}\left((x_1, \ldots, x_n), \bigotimes_{i=1}^{n} \mathcal{S}_i \right) = \bigotimes_{i=1}^{n} \mathcal{N}(x_i, \mathcal{S}_i).$$

6.3 Fixed Topology: Agreement

In this section we study the agreement problem for nonlinear coupled cell systems with fixed topology.

The system under focus is given in (6.4) and the state space $\mathcal{X}$ is set to be $\mathbb{R}^m$. Now some hypotheses on the vector field are assumed. Let

$$\mathcal{C}^i = \text{co}\{x_i, x_j|_{j \in \mathcal{N}_i}\}$$

denote the polytope (convex hull) in $\mathbb{R}^m$ formed by the states of agent i and its neighbors ($\mathcal{N}_i$ is the set of neighbors of agent i, which is fixed also). Then for each $i \in \mathcal{V}$, we assume that

A1': f^i is continuous on $\mathbb{R}^{mn}$;

A2': for all $x \in \mathbb{R}^{mn}$, $f^i(x) \in \mathcal{T}(x_i, \mathcal{C}^i)$. Moreover, $f^i(x) \neq 0$ if x_i is $\mathcal{C}^i$'s vertex and $\mathcal{C}^i$ is not a singleton.

Assumption A1' is to guarantee the existence of solutions. The assumption $f^i(x) \in \mathcal{T}(x_i, \mathcal{C}^i)$, the tangent cone to the set $\mathcal{C}^i$ at x_i, is sometimes referred to as a *sub-tangentiality condition* [23]. Fig. 6.7 illustrates two example situations of A2′. In the left-hand example, agent 1 has only one neighbor, agent 2; the

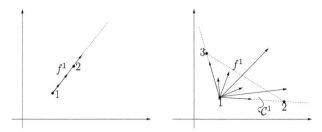

Fig. 6.7. Vector fields satisfying assumption A2′.

polytope $\mathcal{C}^1$ is the line segment joining x_1 and x_2; the tangent cone $\mathcal{T}(x_1, \mathcal{C}^1)$ is the closed ray $\{\lambda(x_2 - x_1) : \lambda \geq 0\}$ (in the picture it's shown translated to x_1); and A2′ means that f^1 is nonzero and points in the direction of $x_2 - x_1$. In the right-hand example, agent 1 has two neighbors, agents 2 and 3; the polytope $\mathcal{C}^1$ is the triangle with vertices x_1, x_2, x_3; the tangent cone $\mathcal{T}(x_1, \mathcal{C}^1)$ is

$$\{\lambda_1(x_2 - x_1) + \lambda_2(x_3 - x_1) : \lambda_1, \lambda_2 \geq 0\}$$

(again, it's shown translated to x_1); and A2′ means that f^1 is nonzero and points into this closed cone. In general, A2′ requires that $f^i(x)$ have the form

$$\sum_{j \in \mathcal{N}_i} \alpha_j(x)(x_j - x_i), \tag{6.5}$$

where $\alpha_j(x)$ are non-negative scalar functions, and that $f^i(x)$ is nonzero if $\mathcal{C}^i$ is not a singleton and x_i is its vertex.

For the nonlinear coupled cell system (6.4) with assumptions A1′ and A2′, it can be checked that Ω is an equilibria set. Let $\bar{x}$ be any point in Ω. Then for any i

$$\mathcal{C}^i = \mathrm{co}\{\bar{x}_i, \bar{x}_j|_{j \in \mathcal{N}_i}\} = \{\bar{x}_i\}.$$

Thus $\mathcal{T}(\bar{x}_i, \mathcal{C}^i) = \{0\}$. It then follows from assumption A2′ that

$$f^i(\bar{x}) = 0, \qquad \text{for all } i = 1, \ldots, n,$$

which implies, Ω is a set of equilibria.

Now we present our main results. The first result is stability property of agreement state. The proof relies on showing that the maximum distance of any agent to a fixed point is nonincreasing.

Let $a \in \mathbb{R}^m$ be any point. Define

$$V_i^a(x) = \frac{1}{2}\|x_i - a\|^2 \quad \text{and} \quad V^a(x) = \max_{i \in \mathcal{V}} V_i(x). \tag{6.6}$$

Lemma 6.2. *A solution $x(t)$ over $[0, \infty)$ exists for the nonlinear coupled cell system (6.4) and along any solution $x(t)$,*

$$D^+ V^a(x(t)) \le 0.$$

Proof: Consider an arbitrary $x^0 \in \mathbb{R}^{mn}$ and let $x(t)$ be a solution of (6.4) defined on the maximal interval $[0, \omega) \subseteq [0, \infty)$ with $x(0) = x^0$. Such a solution exists by Peano's Theorem.

Due to the maximum function in $V^a(x)$, it is not differential everywhere. So we use the Dini derivative here. Define $\mathcal{I}(x) = \{i \in \mathcal{V} : V_i^a(x) = V^a(x)\}$, the set of indices where the maximum is reached. By Lemma B.2, it follows that

$$D^+ V^a(x(t)) = \max_{i \in \mathcal{I}(x(t))} \dot{V}_i^a(x(t)). \tag{6.7}$$

Define a ball in $\mathbb{R}^m$ as

$$\mathcal{B}_a(x) = \left\{ y \in \mathbb{R}^m : \|y - a\|^2 \le 2V^a(x) \right\}.$$

That is, the ball $\mathcal{B}_a(x)$ encloses all the points $x_1, \ldots, x_n$. Then by convexity

$$\mathcal{C}^i = \mathrm{co}\{x_i, x_j|_{j \in \mathcal{N}_i}\} \subset \mathcal{B}_a(x)$$

(see Fig. 6.8, the dashed lines with arrows represent the arcs of the interaction graph). By Lemma 6.1 and assumption A2', we have

$$f^i(x) \in \mathcal{T}(x_i, \mathcal{C}^i) \subset \mathcal{T}(x_i, \mathcal{B}_a(x)).$$

In addition, when $i \in \mathcal{I}(x)$, it means that x_i lies on the boundary of the ball $\mathcal{B}_a(x)$ and so the vector $(x_i - a)$ is a radius of the ball. Hence

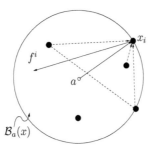

Fig. 6.8. Illustration for $\mathcal{B}_a(x)$ and $f^i(x)$.

$$(x_i - a) \in \mathcal{N}\left(x_i, \mathcal{B}_a(x)\right),$$

the normal cone to $\mathcal{B}_a(x)$ at x_i. Thus, by the definition of normal cone, it follows that for each $i \in \mathcal{I}(x)$,

$$\dot{V}_i^a(x(t)) = (x_i - a)^T f^i(x) \leq 0.$$

Together with (6.7), this leads to

$$D^+ V^a(x(t)) \leq 0.$$

In addition, from above and Lemma B.1, one sees that $V^a(x(t)) \leq V^a(x(0))$ for all $t \in [0, \omega)$ and so $x(t)$ is bounded. Thus, $\omega = \infty$. ∎

Theorem 6.1. *Every agreement state is stable for the nonlinear coupled cell system (6.4).*

Proof: Consider any equilibrium $\bar{x} \in \Omega$. It will be of the form $\bar{x} = \mathbf{1}_n \otimes \zeta$ for some $\zeta \in \mathbb{R}^m$. Let

$$V^\zeta(x) = \frac{1}{2} \max_{i \in \mathcal{V}} \|x_i - \zeta\|^2.$$

It can be easily verified that $V^\zeta(x) = 0$ when $x = \bar{x}$ and $V^\zeta(x) > 0$ when $x \neq \bar{x}$. That is, the function $V^\zeta(x)$ is positive definite with respect to the equilibrium $\bar{x}$.

In addition, by Lemma 6.2, we obtain that the Dini derivative along any trajectory of (6.4)

$$D^+V^\zeta(x(t)) \leq 0.$$

Then by Theorem 6.2 in [114], page 89, it follows that $\bar{x} \in \Omega$ is stable.   ∎

The second result shows the relevance of the interaction graph $\mathcal{G}$ to the problem of achieving state agreement. A necessary and sufficient condition is obtained via nonsmooth analysis with the LaSalle's invariance principle playing a central role.

Theorem 6.2. *The nonlinear coupled cell system (6.4) achieves state agreement globally if and only if the interaction graph $\mathcal{G}$ is quasi-strongly connected.*

Proof: ($\Longleftarrow$) Consider an arbitrary $x^0 \in \mathbb{R}^{mn}$. By Lemma 6.2 we know the system (6.4) has a solution $x(t)$ over $[0, \infty)$ with $x(0) = x^0$. Let $V^a(x)$ be a function of the form (6.6), where $a \in \mathbb{R}^m$ is an arbitrary point. By Lemma 6.2 again, we have

$$D^+V^a(x(t)) \leq 0.$$

Let $\Lambda^+(x^0)$ be the positive limit set of solutions satisfying $x(0) = x^0$. From the proof of Lemma 6.2, we know that $x(t)$ is bounded. Then, by Lemma C.1, the positive limit set $\Lambda^+(x^0)$ is nonempty, compact, and connected. Moreover,

$$x(t) \to \Lambda^+(x^0) \text{ as } t \to \infty.$$

On the other hand, it follows from Theorem C.1 that

$$\Lambda^+(x^0) \subset \mathcal{M},$$

where $\mathcal{M}$ is the union of all solutions that remain in

$$\mathcal{Z}_a = \{x \in \mathbb{R}^{mn} : D^+V^a(x) = 0\}.$$

Choose any two arbitrary points $b, c \in \mathbb{R}^m$. Then by the same argument and Theorem C.1,

$$\Lambda^+(x^0) \subset \mathcal{M}',$$

too, where $\mathcal{M}'$ is the union of all solutions that remain in $\mathcal{Z}_b \cap \mathcal{Z}_c$.

Next, we claim that $\mathcal{M}' \subset \Omega$. To prove this, suppose conversely that there exists a point $q = (q_1 \cdots q_n) \in \mathcal{M}'$ but $q \notin \Omega$. Denote the k vertices ($2 \leq k \leq n$ because $q \notin \Omega$) of co$\{q_1, \ldots, q_n\}$ by $z_1, \ldots, z_k$ (see Fig. 6.9). Since b, c can be

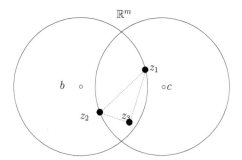

Fig. 6.9. Vertices z_1, z_2, and z_3.

chosen freely, without loss of generality we can assume that there is one and only one vertex, say z_1, in the boundary of $\mathcal{B}_b(q)$ and there is one and only one vertex, say z_2, in the boundary of $\mathcal{B}_c(q)$.

Let

$$\mathcal{I}'(q) = \{i \in \mathcal{V} : q_i = z_1\}$$

and

$$\mathcal{I}''(q) = \{i \in \mathcal{V} : q_i = z_2\}$$

be the sets of agents located at z_1 and z_2, respectively. If the interaction graph $\mathcal{G}$ is quasi-strongly connected, it follows from Theorem 2.1 that there is a center node, say v_c. Since $\mathcal{I}'(q)$ and $\mathcal{I}''(q)$ are disjoint, then the center node v_c cannot be in both sets. Without loss of generality, say it does not belong to $\mathcal{I}'(q)$.

Consider $x(0) = q$ and let $x(t)$ be any solution of (6.4) leaving from q. Since

$$q \in \mathcal{M}' \subset \mathcal{Z}_b \cap \mathcal{Z}_c,$$

we have that, for all $t \in [0, \infty)$,

$$D^+ V^b(x(t)) = \max_{i \in \mathcal{I}'(x(t))} (x_i(t) - b)^T f^i(x(t)) = 0,$$

where

$$\mathcal{I}'(x(t)) = \{i \in \mathcal{V} : V^b(x(t)) = V_i^b(x(t))\}.$$

Notice that we can choose b outside of a compact set containing all $x_i(t), t \in [0, \infty)$, so for all $i \in \mathcal{I}'(x(t))$,

$$x_i(t) - b \neq 0, \quad \forall t \in [0, \infty).$$

Further, by assumption A2' and the fact that $T(x_i, C^i)$ is strictly contained in $T(x_i, \mathcal{B}_b(x))$, $f^i(x(t))$ is not perpendicular to $(x_i(t) - b)$. So

$$D^+ V^b(x(t)) = 0$$

implies that at each t there exists an $i \in \mathcal{I}'(x(t))$ such that $f^i(x(t)) = 0$, i.e.,

$$\dot{x}_i(t) = 0.$$

Next, by the definition of the set $\mathcal{I}'(x(0))$, one has

$$g(x(0)) := \min_{i \in \mathcal{I}'(x(0))} V_i^b(x(0)) - \max_{j \in \mathcal{V} - \mathcal{I}'(x(0))} V_j^b(x(0)) > 0.$$

Since $t \mapsto g(x(t))$ is a continuous function, it follows that there exists a sufficiently small positive scalar ω_1 such that

$$\forall t \in [0, \omega_1], \ g(x(t)) > 0,$$

that is, for any $i \in \mathcal{I}'(x(0))$ and $j \in \mathcal{V} - \mathcal{I}'(x(0))$,

$$V_i^b(x(t)) > V_j^b(x(t)).$$

Hence,

$$(\mathcal{V} - \mathcal{I}'(x(0))) \cap \mathcal{I}'(x(t)) = \phi, \ \forall t \in [0, \omega_1]$$

or, what is the same,

$$\mathcal{I}'(x(t)) \subseteq \mathcal{I}'(x(0)), \ \forall t \in [0, \omega_1].$$

Now partition the set $\mathcal{I}'(x(t))$ as $\mathcal{I}'(x(t)) = \mathcal{J}(x(t)) \cup \bar{\mathcal{J}}(x(t))$, where

$$\mathcal{J}(x(t)) = \{i \in \mathcal{I}'(x(t)) : f^i(x(t)) = 0\}$$
$$\bar{\mathcal{J}}(x(t)) = \{i \in \mathcal{I}'(x(t)) : f^i(x(t)) \neq 0\}.$$

By construction, for all $i \in \mathcal{I}'(x(0))$, $x_i(0) = z_1$. Thus, $i \in \bar{\mathcal{J}}(x(0))$ implies

$$x_i(0) = z_1 \text{ and } f^i(x(0)) \neq 0,$$

which in turn implies that agent i has a neighbor in $\mathcal{V} - \mathcal{I}'(x(0))$. Indeed, if all neighbors of $i \in \bar{\mathcal{J}}(x(0))$ were in $\mathcal{I}'(x(0))$, then $\mathcal{C}^i(x(0))$ would be a singleton and hence necessarily, from assumption A2', $f^i(x(0)) = 0$, contradicting the fact that $i \in \bar{\mathcal{J}}(x(0))$.

Let $j \in \mathcal{V} - \mathcal{I}'(x(0))$ be a neighbor agent of $i \in \bar{\mathcal{J}}(x(0))$. Since, for all $t \in [0, \omega_1]$,

$$\mathcal{I}'(x(t)) \subseteq \mathcal{I}'(x(0)),$$

it follows that

$$\mathcal{V} - \mathcal{I}'(x(0)) \subseteq \mathcal{V} - \mathcal{I}'(x(t)),$$

and hence

$$j \in \mathcal{V} - \mathcal{I}'(x(t)) \text{ for all } t \in [0, \omega_1].$$

If, for all $t \in [0, \omega_1]$, $i \in \mathcal{I}'(x(t))$ implies that $x_i(t)$ is on the boundary of the ball $\mathcal{B}_b(x(t))$ and so $x_i(t)$ is a vertex of $\mathcal{C}^i(x(t))$. Moreover, since agent i has a neighbor $j \in \mathcal{V} - \mathcal{I}'(x(t))$, $\mathcal{C}^i(x(t))$ is not a singleton. By assumption A2' one has that $f^i(x(t)) \neq 0$. We have thus shown that

$$\big(\forall t \in [0, \omega_1]\big) \ i \in \mathcal{I}'(x(t)) \implies \big(\forall t \in [0, \omega_1]\big) \ f^i(x(t)) \neq 0.$$

This in particular implies that, during the interval $[0, \omega_1]$, no agent in $\bar{\mathcal{J}}(x(0))$ can get to $\mathcal{J}(x(t))$ or, equivalently,

$$\big(\forall t \in [0, \omega_1]\big) \ \mathcal{J}(x(t)) \subseteq \mathcal{J}(x(0)).$$

Next, we show that $\mathcal{J}(x(t))$ is strictly contained in $\mathcal{J}(x(0))$. Suppose, by way of contradiction, that $\mathcal{J}(x(t)) = \mathcal{J}(x(0))$ for all $t \in (0, \omega_1]$. Then

$$\big(\forall i \in J(x(0))\big) \ \big(\forall t \in [0, \omega_1]\big) \ f^i(x(t)) = 0.$$

This means that, during $[0, \omega_1]$, $x_i(t) = z_1$ for all $i \in \mathcal{J}(x(0))$. Since $x_i(t)$ is a vertex of $\mathcal{C}^i(x(t))$ and $f^i(x(t)) = 0$, assumption A2' implies that $\mathcal{C}^i(x(t))$ is a singleton. Hence, for each $i \in \mathcal{J}(x(0))$ and all $t \in [0, \omega_1]$, all neighbor agents of agent i are collocated at z_1. Since the center node of $\mathcal{G}$ does not belong to $\mathcal{J}(x(0)) \subset \mathcal{I}'(x(0))$, some agent in $\mathcal{J}(x(0))$ must have a neighbor agent j in $\bar{\mathcal{J}}(x(0)) \cup \mathcal{V} - \mathcal{I}'(x(0))$. As shown above, such a neighbor agent j is such that $x_j(t) = z_1$ for all $t \in [0, \omega_1]$ which implies that $j \notin \mathcal{V} - \mathcal{I}'(x(0))$. On the other hand, the fact that $\forall t \in [0, \omega_1]$, $x_j(t) = z_1$, also implies that $f^j(x(0)) = 0$ contradicting the fact that $j \in \bar{\mathcal{J}}(x(0))$.

We have thus shown that there exists $t_1 \in (0, \omega_1]$ such that $\mathcal{J}(x(t_1))$ is a strict subset of $\mathcal{J}(x(0))$. A repetition of this argument leads to the existence of t_k such that $\mathcal{J}(x(t_k))$ is empty, which contradicts the fact that there exists an $i \in \mathcal{I}'(x(t))$ such that $f^i(x(t)) = 0$.

Thus, the solution $x(t) \to \Omega$ as $t \to \infty$ and the system achieves state agreement.

($\Longrightarrow$) To prove the contrapositive form, assume that $\mathcal{G}$ is not quasi-strongly connected, that is, there are two nodes i^* and j^* such that for any node k, either i^* or j^* is not reachable from k. Let $\mathcal{V}_1$ be the subset of nodes from which i^* is reachable and $\mathcal{V}_2$ be the subset of nodes from which j^* is reachable. Obviously, $\mathcal{V}_1$ and $\mathcal{V}_2$ are disjoint. Moreover, for each node $i \in \mathcal{V}_1$ (resp. $\mathcal{V}_2$), the set of neighbors of node i is a subset of $\mathcal{V}_1$ (resp. $\mathcal{V}_2$).

Choose any $z_1, z_2 \in \mathbb{R}^m$ such that $z_1 \neq z_2$, and pick initial conditions

$$x_i(0) = \begin{cases} z_1, & \forall\, i \in \mathcal{V}_1, \\ z_2, & \forall\, i \in \mathcal{V}_2. \end{cases}$$

Then by assumption A2′,

$$x_i(t) = \begin{cases} z_1, & \forall\, i \in \mathcal{V}_1, \\ z_2, & \forall\, i \in \mathcal{V}_2, \end{cases} \quad \forall\, t \geq 0.$$

This proves that the system does not achieve state agreement globally.   ∎

6.4 Dynamic Topology: Set Invariance and Stability

In this section we turn our attention to the nonlinear coupled cell system with dynamic topology. The mathematical models are given in (6.1)–(6.3). Now some hypotheses are introduced on the general vector fields of the family (6.1).

Let

$$\mathcal{C}_p^i = \mathrm{co}\{x_i, x_j|_{j \in \mathcal{N}_i(p)}\}$$

denote the polytope formed by the states of agent i and its neighbors. Also, it is assumed that all the agent's states lie in the common state space, $\mathcal{X} \subset \mathbb{R}^m$, that plays the role of a region of focus. Thus, agreement will occur in $\mathcal{X}$. We assume that, for each $i \in \mathcal{V}$ and each $p \in \mathcal{P}$, the vector fields $f_p^i : \mathcal{X}^n \to \mathcal{X}$ satisfy the following two assumptions:

A1: f_p^i is locally Lipschitz on $\mathcal{X}^n$;

A2: For all $x \in \mathcal{X}^n$, $f_p^i(x) \in \mathrm{ri}\left(\mathcal{T}(x_i, \mathcal{C}_p^i)\right)$.

Assumption A2 is sometimes referred to as a *strict sub-tangentiality condition*. Compared with assumption A2′ that we introduced in the previous section, assumption A2 excludes the case where $f_p^i(x)$, viewed as a vector applied at the point x_i, is tangent to the relative boundary of the convex set $\mathcal{C}_p^i$. Two example situations of A2 are illustrated in Fig. 6.10.

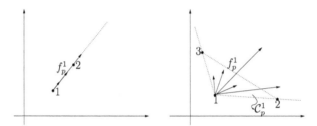

Fig. 6.10. Vector fields satisfying assumption A2.

For the coupled linear system (5.1) we studied in Chapter 5, a comparison with (6.5) makes it clear that for each $i \in \mathcal{V}$ and each $p \in \mathcal{P}$, the vector fields in (5.1) satisfy assumptions A1 and A2 with $\mathcal{X} = \mathbb{R}^m$ (and of course satisfy A1′ and A2′).

Similar to the linear case, we let $\mathcal{S}_{dwell}(\tau_D)$ denote the class of piecewise constant switching signals with dwell time τ_D and make the following assumption:

A3: $\sigma(t) \in \mathcal{S}_{dwell}(\tau_D)$.

Now, we introduce an important result for positive invariance property of any compact convex set. It is perhaps helpful to give some intuition using a 2D example. For $m = 2$, all agents move in the plane. Let $\mathcal{A}$ be a compact convex subset of $\mathcal{X} \subset \mathbb{R}^2$ and assume all agents start in $\mathcal{A}$. Let $\mathcal{C}(t)$ denote the convex hull of the agents' locations at time t. Because $\mathcal{A}$ is convex, clearly $\mathcal{C}(0) \subset \mathcal{A}$. Now invoke assumption A2. An agent that is initially in the interior of $\mathcal{C}(0)$ can head off in any direction at $t = 0$, but an agent that is initially on the boundary of $\mathcal{C}(0)$ is constrained to head into its interior. In this way, $\mathcal{C}(t)$ is non-increasing (if $t_2 > t_1$, then $\mathcal{C}(t_2) \subseteq \mathcal{C}(t_1)$). Saying that the agents cannot exit the set $\mathcal{A}$ is mathematically equivalent to saying that the Cartesian product $\mathcal{A}^n$ is positively invariant for the switched nonlinear coupled cell system (6.3).

Theorem 6.3. *Let $\mathcal{A} \subset \mathcal{X}$ be a compact convex set. Then $\mathcal{A}^n$ is positively invariant for the switched nonlinear coupled cell system (6.3).*

The proof requires Nagumo's theorem concerning set invariance.

Theorem 6.4 ([8], Nagumo, 1942). *Consider the system $\dot{y} = F(y)$, with $F : \mathbb{R}^l \to \mathbb{R}^l$, and let $\mathcal{Y} \subset \mathbb{R}^l$ be a closed convex set. Assume that, for each y^0 in $\mathcal{Y}$, there exists $\epsilon(y^0) > 0$ such that the system admits a unique solution $y(t, y^0)$ defined for all $t \in [0, \epsilon(y^0))$. Then,*

$$y^0 \in \mathcal{Y} \implies y(t, y^0) \in \mathcal{Y}, \quad \forall t \in [0, \epsilon(y^0))$$

if and only if $F(y) \in T(y, \mathcal{Y})$ for all $y \in \mathcal{Y}$.

Proof of Theorem 6.3: Let $\mathcal{A}$ be any compact convex set in $\mathcal{X}$ and consider any initial state $x^0 \in \mathcal{A}^n$ and any initial time t_0. For any switching signal $\sigma(t) \in \mathcal{S}_{dwell}(\tau_D)$, let $x(t, t_0, x^0)$ be the solution of the switched system (6.3) with $x(t_0) = x^0$, and let $[t_0, t_0 + \epsilon(t_0, x^0))$ be its maximal interval of existence.

For any point $x \in \mathcal{A}^n$, it is obvious that $\mathcal{C}_p^i \subset \mathcal{A}$ for all $i \in \mathcal{V}$ and $p \in \mathcal{P}$, by convexity of $\mathcal{A}$. Thus, by property (a) in Lemma 6.1,

$$f_p^i(x) \in \mathrm{ri}\left(T(x_i, \mathcal{C}_p^i)\right) \subset T(x_i, \mathcal{A}), \quad \forall i \in \mathcal{V}, \ \forall p \in \mathcal{P},$$

and by property (b) in the same lemma,

$$g(t, x) := f_{\sigma(t)}(x) \in T(x, \mathcal{A}^n) \text{ for all } t \in \mathbb{R} \text{ and } x \in \mathcal{A}^n.$$

Set $y = (t, x)$ and construct the augmented system

$$\dot{y} = F(y) := \begin{pmatrix} 1 \\ g(y) \end{pmatrix}. \tag{6.8}$$

Since $g(t, x)$ admits a unique solution $x(t, t_0, x^0)$ defined for all $t \in [t_0, t_0 + \epsilon(t_0, x^0))$, it follows that for all $y^0 = (t_0, x^0) \in \mathbb{R} \times \mathcal{A}^n$, the augmented system (6.8) has a unique solution $y(t, y^0)$ defined on $[0, \epsilon(y^0))$. Moreover,

$$F(y) \in T(t, \mathbb{R}) \times T(x, \mathcal{A}^n) = T(y, \mathbb{R} \times \mathcal{A}^n) \text{ for all } y \in \mathbb{R} \times \mathcal{A}^n.$$

Since $\mathbb{R} \times \mathcal{A}^n$ is closed and convex, by Theorem 6.4 (Nagumo's Theorem) it follows that

$$y^0 = (t_0, x^0) \in \mathbb{R} \times \mathcal{A}^n \implies y(\tau) \in \mathbb{R} \times \mathcal{A}^n, \; \forall \tau \in \big[0, \epsilon(y^0)\big). \qquad (6.9)$$

The solution $y(\tau)$ to (6.8) with initial condition $y^0 = (t_0, x^0)$ is related to the solution $x(t)$ to $\dot{x} = g(t, x)$ with initial condition $x(t_0) = x^0$ as follows

$$\big(t, x(t)\big) = y(t - t_0), \; \forall t \in \big[t_0, t_0 + \epsilon(t_0, x^0)\big).$$

We thus rewrite condition (6.9) as

$$t_0 \in \mathbb{R} \text{ and } x^0 \in \mathcal{A}^n \implies x(t) \in \mathcal{A}^n, \; \forall t \in \big[t_0, t_0 + \epsilon(t_0, x^0)\big).$$

Since the set $\mathcal{A}^n$ is compact, it follows by Theorem 2.4 in [69] that, for all $x^0 \in \mathcal{A}^n$ and all t_0, $\epsilon(t_0, x^0) = \infty$ and the set $\mathcal{A}^n$ is positively invariant for the switched interconnected system (6.3). ■

The second result establishes uniform stability property of agreement state.

Theorem 6.5. *Every agreement state is uniformly stable for the switched nonlinear coupled cell system (6.3).*

Proof: For any equilibrium $\bar{x} \in \Omega$, it must be of the form

$$\bar{x} = \mathbf{1}_n \otimes \zeta,$$

where $\zeta \in \text{int}(\mathcal{X})$. Let $\varepsilon > 0$ be arbitrary. We choose δ $(0 < \delta \leq \varepsilon)$ small enough so that the box

$$\mathcal{A}_\delta(\zeta) := \Big\{ y \in \mathbb{R}^m : \|y - \zeta\|_\infty \leq \delta \Big\}$$

is still within $\mathcal{X}$. Obviously, $\mathcal{A}_\delta(\zeta)$ is a compact convex set. It follows from Theorem 6.3 that the Cartesian product $\mathcal{A}_\delta^n(\zeta)$ is positively invariant for the system (6.3).

Notice that $x \in \mathcal{A}_\delta^n(\zeta)$ is equivalent to

$$\|x - \bar{x}\|_\infty \leq \delta$$

and also

$$\mathcal{A}_\delta^n(\zeta) \subseteq \mathcal{A}_\varepsilon^n(\zeta).$$

We have thus proven that $\forall \varepsilon > 0$, $\exists \delta > 0$ such that $\forall t_0$

$$\|x^0 - \bar{x}\| \leq \delta \Longrightarrow (\forall t \geq t_0) \ \|x(t) - \bar{x}\| \leq \varepsilon.$$

Hence, the conclusion follows.   ∎

6.5 Dynamic Topology: Agreement

This section continues to study the switched nonlinear coupled cell system (6.3) with assumptions A1–A3, but focuses on the problem of finding necessary and sufficient conditions so that the system achieves state agreement.

Theorem 6.6. *Suppose $\mathcal{X}$ is closed and convex. The switched nonlinear coupled cell system (6.3) with assumptions A1–A3 achieves state agreement in $\mathcal{X}$ uniformly if and only if the interaction graph $\mathcal{G}(t)$ is uniformly quasi-strongly connected.*

If assumptions A1 and A2 hold $\mathbb{R}^m$, then the switched nonlinear coupled cell system (6.3) globally uniformly achieves state agreement if and only if $\mathcal{G}(t)$ is uniformly quasi-strongly connected.

Now we are going to prove Theorem 6.6. Before proceeding, we need some lemmas. The first three are basic properties of continuity, Lipschitz continuity, and class $\mathcal{KL}$ function, respectively. The remaining lemmas establish some technical properties that hold for the system (6.3).

Lemma 6.3. *Suppose that $f : \mathbb{R}^l \times \mathcal{M} \to \mathbb{R}$ is locally Lipschitz in its first argument, where $\mathcal{M}$ is a compact subset of $\mathbb{R}^k$. Then*

$$g : \mathbb{R}^l \to \mathbb{R}, \ y \mapsto \max \big\{ f(y, z) : \ z \in \mathcal{M} \big\}$$

is locally Lipschitz in y.

Proof: Since f is locally Lipschitz in y, then for each $z \in \mathcal{M}$, there exists a constant L_z such that $\forall y, y' \in \mathcal{B}_r(y_0) := \big\{ y \in \mathbb{R}^l : \ \|y - y_0\| \leq r \big\}$,

$$\|f(y, z) - f(y', z)\| \leq L_z \|y - y'\|.$$

On the other hand, for all $y, y' \in \mathcal{B}_r(y_0)$,

$$\|g(y) - g(y')\| = \|\max_{z \in \mathcal{M}} f(y, z) - \max_{z \in \mathcal{M}} f(y', z)\|.$$

Let

$$\max_{z \in \mathcal{M}} f(y, z) = f(y, z_y)$$

and

$$\max_{z \in \mathcal{M}} f(y', z) = f(y', z_{y'}).$$

Then

$$f(y, z_y) \geq f(y, z_{y'}) \quad \text{and} \quad f(y', z_{y'}) \geq f(y', z_y).$$

So there exists at least one λ satisfying $0 \leq \lambda \leq 1$ (see Fig. 6.11) such that when

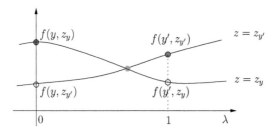

Fig. 6.11. Intersecting point.

$\bar{y} = (1 - \lambda)y + \lambda y'$,

$$f(\bar{y}, z_y) = f(\bar{y}, z_{y'}).$$

Thus,

$$
\begin{aligned}
\|g(y) - g(y')\| &= \|f(y, z_y) - f(\bar{y}, z_y) + f(\bar{y}, z_{y'}) - f(y', z_{y'})\| \\
&\leq \|f(y, z_y) - f(\bar{y}, z_y)\| + \|f(\bar{y}, z_{y'}) - f(y', z_{y'})\| \\
&\leq L_{z_y} \|y - \bar{y}\| + L_{z_{y'}} \|\bar{y} - y'\| \\
&= \lambda L_{z_y} \|y - y'\| + (1 - \lambda) L_{z_{y'}} \|y - y'\|.
\end{aligned}
$$

Set $L = \max_{z \in \mathcal{M}} L_z$. Then it follows that

$$\forall y, y' \in \mathcal{B}_r(y_0), \ \|g(y) - g(y')\| \leq L\|y - y'\|,$$

which shows the function $g(\cdot)$ is locally Lipschitz in its argument. ∎

Lemma 6.4. *Let $f : \mathbb{R}^m \to \mathbb{R}$ be continuous and, given $\xi \in \mathbb{R}^l$ and $A \in \mathbb{R}^{l \times m}$, assume that the set $\{z \in \mathbb{R}^m : Az \preceq \xi\}$ is compact (the inequality is imposed* *component-wise). Then,*

$$g : \mathbb{R}^l \to \mathbb{R}$$
$$\xi \mapsto \min \left\{ f(z) : \; Az \preceq \xi \right\},$$

is continuous.

The lemma is straightforward from Theorem E.1 (Berge's Maximum Theorem).

Lemma 6.5. *Let* $g : [0, a) \to [0, \infty)$ *be a locally Lipschitz and positive definite function, where* a *is a positive real number. Then, for all* $y^0 \in [0, a)$, *the differential equation*

$$\dot{y} = -g(y), \quad y(t_0) = y^0$$

has a unique solution $y(t) = \gamma(y^0, t - t_0)$ *defined for all* $t \geq t_0$, *where*

$$\gamma : [0, a) \times [0, \infty) \to [0, \infty)$$

is a class $\mathcal{KL}$ *function.*

The proof of the lemma above employs the same arguments of the proof of Lemma 3.4 in [69].

Now we are ready to establish certain technical properties of the system (6.3). First, we introduce some notations that are used in the following proofs.

Define a hyper-cube in $\mathbb{R}^m$ by

$$\mathcal{A}_r(z) = \left\{ y \in \mathbb{R}^m : \|y - z\|_\infty \leq r \right\}.$$

Let $c > 0$ be large enough that $\mathcal{X}_c := \mathcal{X} \cap \mathcal{A}_c(0)$ is not empty. For any $x = (x_1, \ldots, x_n) \in \mathcal{X}_c^n$, the Cartesian product of n copies of $\mathcal{X}_c$, we let

$$\mathcal{C}(x) = \mathrm{co}\{x_1, \ldots, x_n\},$$

the polytope of $x_1, \ldots, x_n$. In addition, for each $j = 1, \ldots, m$, we let

$$a_j(x) = \max_{i \in \mathcal{V}} x_{ij} \quad \text{and} \quad b_j(x) = \min_{i \in \mathcal{V}} x_{ij}, \tag{6.10}$$

where x_{ij} is the jth entry of $x_i \in \mathbb{R}^m$. In what follows, we call the set

$$\{y \in \mathcal{C}(x) : y_1 = a_1(x)\}$$

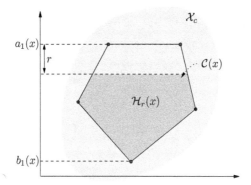

Fig. 6.12. Notations.

the *first upper boundary* of $\mathcal{C}(x)$. Finally, for small enough $r > 0$, we define

$$\mathcal{H}_r(x) = \{y \in \mathcal{C}(x) : y_1 \leq a_1(x) - r\}.$$

See Fig. 6.12 for an example.

In the following two lemmas, we assume that the hypotheses of Theorem 6.6 hold. That is, $\mathcal{X}$ is a closed convex set and the interaction graph $\mathcal{G}(t)$ is uniformly quasi-strongly connected.

Remark 6.1. If the interaction graph $\mathcal{G}(t)$ is uniformly quasi-strongly connected, it can be easily seen that there is some $p \in \mathcal{P}$ such that the mode graph $\mathcal{G}_p$ has a nonempty arc set $\mathcal{E}_p$.

The first lemma shows that if at time t' all agents are in $\mathcal{X}_c$, then the agents that at some time $t_1 \geq t'$ are in the interior of $\mathcal{C}(x(t'))$ cannot reach the first upper boundary of $\mathcal{C}(x(t'))$ in finite time. See Fig. 6.13 for an illustration.

Lemma 6.6. *Given $c > 0$ large enough that $\mathcal{X}_c \neq \phi$, there exists a class $\mathcal{KL}$ function $\gamma : [0, 2c] \times [0, \infty) \rightarrow [0, \infty)$ with the property $\gamma(\Delta, 0) = \Delta$ such that the following holds: For any $(t', x(t')) \in \mathbb{R} \times \mathcal{X}_c^n$, any $\varepsilon > 0$ sufficiently small, and any $T > 0$, if there exists an i such that $x_i(t_1) \in \mathcal{H}_\varepsilon(x(t'))$ at some $t_1 \geq t'$, then $x_i(t) \in \mathcal{H}_\delta(x(t'))$ for all $t \in [t_1, t_1 + T]$ with $\delta = \gamma(\varepsilon, T)$.*

Proof: For an arbitrarily large $c > 0$, we consider any $(t', x(t')) \in \mathbb{R} \times \mathcal{X}_c^n$ and let $\varsigma = x(t')$. Then we define

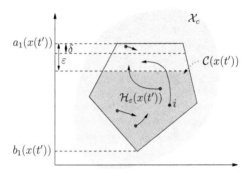

Fig. 6.13. Illustration for Lemma 6.6.

$$\bar{f}_\varsigma : \left[b_1(\varsigma),\ a_1(\varsigma) \right] \to \mathbb{R}$$

$$w \mapsto \max_{i \in \mathcal{V}} \left\{ \max \left\{ f_p^{i1}(x) : p \in \mathcal{P},\ x \in \mathcal{C}^n(\varsigma)\ \text{such that}\ x_{i1} = w \right\} \right\},$$

where f_p^{i1} and x_{i1} are the first component of f_p^i and x_i, respectively. By assumption A1, f_p^i is locally Lipschitz, so it follows from Lemma 6.3 that $\bar{f}_\varsigma$ is locally Lipschitz on its domain.

Next, we are going to show that

$$\begin{cases} \bar{f}_\varsigma(w) = 0 & \text{if } w = a_1(\varsigma) \\ \bar{f}_\varsigma(w) > 0 & \text{if } w \in \big[b_1(\varsigma),\ a_1(\varsigma) \big). \end{cases}$$

Firstly, for each $i \in \mathcal{V}$ and each $p \in \mathcal{P}$, notice that $x \in \mathcal{C}^n(\varsigma)$ and $x_{i1} = a_1(\varsigma)$ imply that $\mathcal{C}_p^i \subset \mathcal{C}(\varsigma)$ and x_i is on the first upper boundary of $\mathcal{C}(\varsigma)$. By assumption A2 and Lemma 6.1, $f_p^i(x) \in \mathrm{ri}\left(\mathcal{T}(x_i, \mathcal{C}_p^i) \right) \subset \mathcal{T}\left(x_i, \mathcal{C}(\varsigma) \right)$, and so $f_p^{i1}(x) \le 0$. Next choose $i \in \mathcal{V}$ and $p \in \mathcal{P}$ such that $\mathcal{E}_p$ is non-empty, that is, agent i has at least one neighbor agent. Such a pair (i, p) exists by Remark 6.1. Pick $x \in \mathcal{C}^n(\varsigma)$ so that $x_{i1} = a_1(\varsigma)$ and, for all $j \in \mathcal{N}_i(p)$, $x_j = x_i$. Since $\mathcal{C}_p^i$ is the singleton $\{x_i\}$, it follows from assumption A2 that $f_p^i(x) = 0$, implying $\bar{f}_\varsigma\left(a_1(\varsigma) \right) = 0$. Secondly, for each $w \in \big[b_1(\varsigma),\ a_1(\varsigma) \big)$, choose (i, p) as before and pick $x \in \mathcal{C}^n(\varsigma)$ such that $x_{i1} = w$ and, for all $j \in \mathcal{N}_i(p)$, $x_{j1} = a_1(\varsigma)$ and $x_{jk} = x_{ik}$, $k = 2, \ldots, m$. Now $\mathcal{C}_p^i$ is the line segment with vertices $(a_1(\varsigma), x_{i2}, \ldots, x_{im})$ and $(w, x_{i2}, \ldots, x_{im})$, so by assumption A2 $f_p^{i1}(x) > 0$, implying that $\bar{f}_\varsigma(w) > 0$.

Letting $y := a_1(\varsigma) - w$, we define

$$h_\varsigma(y) := \bar{f}_\varsigma(a_1(\varsigma) - y).$$

Then it is easily followed that the function $h_\varsigma(\cdot)$ is locally Lipschitz and is positive definite on $[0, a_1(\varsigma) - b_1(\varsigma)]$. Extend its domain to $[0, 2c]$ by the following construction

$$\bar{h}_\varsigma(y) = \begin{cases} h_\varsigma(y), & y \in \big[0, a_1(\varsigma) - b_1(\varsigma)\big], \\ h_\varsigma\big(a_1(\varsigma) - b_1(\varsigma)\big), & y \in \big(a_1(\varsigma) - b_1(\varsigma), 2c\big]. \end{cases}$$

This function is still locally Lipschitz on y and positive definite. On the other hand, by Lemma 6.4 the function $\bar{h}_\varsigma(y)$ is continuous with respect to the parameter ς. Furthermore, notice that $\mathcal{X}_c^n$ is a compact set, so the function

$$\bar{h}(y) := \max\left\{\bar{h}_\varsigma(y) : \varsigma \in \mathcal{X}_c^n\right\} \tag{6.11}$$

is well-defined, locally Lipschitz by Lemma 6.3, and positive definite. Given any initial condition $(t_0, y^0) \in \mathbb{R} \times [0, 2c]$, the solution of $\dot{y} = -\bar{h}(y)$ is given by $\gamma(y^0, t - t_0)$, which is a class $\mathcal{KL}$ function by Lemma 6.5 and satisfies the property $\gamma(\triangle, 0) = \triangle$.

Consider now any ε sufficiently small and any $T > 0$. If there exists an i such that $x_i(t_1) \in \mathcal{H}_\varepsilon(x(t'))$ at some $t_1 \geq t'$, by recalling $\varsigma = x(t')$ it follows from Theorem 6.3 that $x(t) \in \mathcal{C}^n(\varsigma)$ for all $t \geq t'$ and then clearly it is also true for all $t \geq t_1$. Hence, from the definition of $\bar{f}_\varsigma(\cdot)$ we know that

$$\dot{x}_{i1}(t) = f_{\sigma(t)}^{i1}(x(t)) \leq \bar{f}_\varsigma(x_{i1}(t)) \quad \text{for all } t \geq t_1.$$

Let $w(t)$ be the solution of $\dot{w} = \bar{f}_\varsigma(w)$ with the initial condition $w(t_1) = x_{i1}(t_1)$. Thus, by the Comparison Lemma [69], one obtains $x_{i1}(t) \leq w(t)$ for all $t \geq t_1$.

On the other hand, considering the coordinate transformation $y = a_1(\varsigma) - w$, we know that $y(t) = a_1(\varsigma) - w(t)$ is the solution of

$$\dot{y} = -h_\varsigma(y)$$

and that it is also the solution of

$$\dot{y} = -\bar{h}_\varsigma(y)$$

since the initial condition

$$y(t_1) = a_1(\varsigma) - w(t_1) = a_1(\varsigma) - x_{i1}(t_1)$$

is in $\left[0, a_1(\varsigma) - b_1(\varsigma)\right]$. Furthermore, from (6.11) we know $\bar{h}_\varsigma(y) \le \bar{h}(y)$. Applying the Comparison Lemma [69] again leads to

$$y(t) \ge \gamma(y(t_1), t - t_1) \quad \text{for all } t \ge t_1.$$

Hence, the following inequalities hold for all $t \ge t_1$

$$
\begin{aligned}
x_{i1}(t) \le w(t) &= a_1(\varsigma) - y(t) \le a_1(\varsigma) - \gamma(y(t_1), t - t_1) \\
&= a_1(\varsigma) - \gamma(a_1(\varsigma) - x_{i1}(t_1), t - t_1).
\end{aligned}
$$

Also, $x_i(t_1) \in \mathcal{H}_\varepsilon(x(t'))$ implies

$$a_1(\varsigma) - x_{i1}(t_1) \ge \varepsilon.$$

This together with the fact that $\gamma(\cdot, \cdot)$ is of class $\mathcal{KL}$ lead to that for all $t \in [t_1, t_1 + T]$

$$x_{i1}(t) \le a_1(\varsigma) - \gamma(a_1(\varsigma) - x_{i1}(t_1), t - t_0) \le a_1(\varsigma) - \gamma(\varepsilon, T).$$

It in turn implies that for all $t \in [t_1, t_1 + T]$

$$x_i(t) \in \mathcal{H}_\delta(x(t')) \quad \text{with } \delta = \gamma(\varepsilon, T).$$

■

The next lemma shows that if at time t' all agents are in $\mathcal{X}_c$, then the agents that at some time $t_1 \ge t'$ have a neighbor in the interior of $\mathcal{C}(x(t'))$ for certain amount of time will be in the interior of $\mathcal{C}(x(t'))$ at some time $t_2 \ge t'$. See Fig. 6.14 for an illustration (the dashed line with arrow indicates agent j is a neighbor of agent i in the picture).

Lemma 6.7. *Given $c > 0$ large enough that $\mathcal{X}_c \ne \phi$, there exists a class $\mathcal{K}$ function $\varphi : [0, 2c] \to [0, \infty)$ with the property $\varphi(\triangle) < \triangle$ when $\triangle \ne 0$ such that the following holds: For any $(t', x(t')) \in \mathbb{R} \times \mathcal{X}_c^n$ and any $\delta > 0$ sufficiently small, if there exist a pair (i, j) and a $t_1 \ge t'$ such that $j \in \mathcal{N}_i(t)$ and $x_j(t) \in \mathcal{H}_\delta(x(t'))$ for all $t \in [t_1, t_1 + \tau_D]$, then there exists a $t_2 \in [t', t_1 + \tau_D]$ such that $x_i(t_2) \in \mathcal{H}_\varepsilon(x(t'))$ with $\varepsilon = \varphi(\delta)$.*

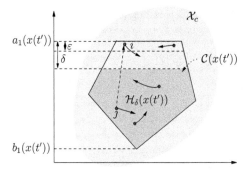

Fig. 6.14. Illustration for Lemma 6.7.

Proof: For an arbitrarily large $c > 0$, we consider any $(t', x(t')) \in \mathbb{R} \times \mathcal{X}_c^n$ and let $\varsigma = x(t')$. For any $\delta > 0$ sufficiently small, and $i \in \mathcal{V}$, $p \in \mathcal{P}$, we define a set

$$\mathcal{O}_\varsigma(i, p, \delta) := \Big\{ x \in \mathcal{C}^n(\varsigma) : \; x_{i1} = a_1(\varsigma) \text{ and } \big(\exists j \in \mathcal{N}_i(p)\big) x_j \in \mathcal{H}_\delta(\varsigma) \Big\}.$$

This is the set of states such that x_i is on the first upper boundary of $\mathcal{C}(\varsigma)$ and at least one of its neighbors (in the digraph $\mathcal{G}_p$), say agent j, has its state x_j in $\mathcal{H}_\delta(\varsigma)$. See Fig. 6.15 for an illustration (the dashed arrows indicate which agents are neighbors of agent i in graph $\mathcal{G}_p$).

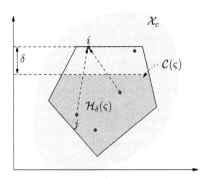

Fig. 6.15. Possible element in $\mathcal{O}_\varsigma(i, p, \delta)$.

Now define the minimum speed of any agent along the first direction when $x \in \mathcal{O}_\varsigma(i, p, \delta)$ and the pair (i, p) ranges over $\mathcal{V} \times \mathcal{P}$,

$$d_\varsigma(\delta) := \min \left\{ |f_p^{i1}(x)| : \ i \in \mathcal{V}, \ p \in \mathcal{P}, \ x \in \mathcal{O}_\varsigma(i, p, \delta) \right\}.$$

Obviously, $d_\varsigma(\delta) \geq 0$. Furthermore, if $\delta > 0$, then by Remark 6.1, there exist a pair $(i, p) \in \mathcal{V} \times \mathcal{P}$ such that $\mathcal{O}_\varsigma(i, p, \delta)$ is non-empty. For any such pair (i, p), assumption A2 implies that, for all $x \in \mathcal{O}_\varsigma(i, p, \delta)$, $|f_p^{i1}(x)| > 0$. It readily follows that $d_\varsigma(\delta) > 0$ when $\delta > 0$.

Notice that $\mathcal{X}_c^n$ is compact and that $f_p(x)$ is locally Lipschitz for each $p \in \mathcal{P}$. So there is a Lipschitz constant L_p such that

$$\forall (x, z) \in \mathcal{X}_c^n \times \mathcal{X}_c^n, \ \|f_p(x) - f_p(z)\| \leq L_p \|x - z\|.$$

Let $L = \max \left\{ L_p : \ p \in \mathcal{P} \right\}$. Then

$$\|f_p(x) - f_p(z)\| \leq L \|x - z\|$$

for all $p \in \mathcal{P}$ and for all $(x, z) \in \mathcal{X}_c^n \times \mathcal{X}_c^n$. Define

$$\varphi_\varsigma(\delta) := \min \left\{ \delta, \ \frac{\tau_D d_\varsigma(\delta)}{\tau_D L + 1} \right\}.$$

We thus know that $\varphi_\varsigma(\delta) = 0$ if $\delta = 0$ and that $\varphi_\varsigma(\delta) > 0$ if $\delta > 0$. In addition, by Lemma 6.4, $\varphi_\varsigma(\delta)$ is continuous on both δ and ς. Extend its domain to $[0, 2c]$ by the following construction

$$\bar{\varphi}_\varsigma(\delta) = \begin{cases} \varphi_\varsigma(\delta), & \delta \in \left[0, a_1(\varsigma) - b_1(\varsigma)\right], \\ \varphi_\varsigma(a_1(\varsigma) - b_1(\varsigma)), & \delta \in \left(a_1(\varsigma) - b_1(\varsigma), 2c\right]. \end{cases}$$

Clearly, this function is also continuous on ς. Since $\mathcal{X}_c^n$ is compact, the function

$$\bar{\varphi}(\delta) := \min \left\{ \bar{\varphi}_\varsigma(\delta) : \varsigma \in \mathcal{X}_c^n \text{ such that } a_1(\varsigma) - b_1(\varsigma) \geq \delta \right\}$$

is well-defined. Furthermore, it is also continuous and positive definite. Hence, there exists a class $\mathcal{K}$ function $\varphi : [0, 2c] \rightarrow [0, \infty)$ such that $\varphi(\delta) < \bar{\varphi}(\delta) \leq \delta$ if $\delta \neq 0$.

Next we are going to show if there exist a pair (i, j) and a $t_1 \geq t'$ such that $j \in \mathcal{N}_i(t)$ and $x_j(t) \in \mathcal{H}_\delta(x(t'))$ for all $t \in [t_1, t_1 + \tau_D]$, then there exists a $t_2 \in [t', t_1 + \tau_D]$ such that $x_i(t_2) \in \mathcal{H}_\varepsilon(x(t'))$ with $\varepsilon = \varphi(\delta)$. Suppose by contradiction that

$$x_i(t) \notin \mathcal{H}_\varepsilon(x(t')) \quad \text{for all } t \in [t', t_1 + \tau_D]. \tag{6.12}$$

On the other hand, from Theorem 6.3 we know $x_i(t) \in \mathcal{C}(x(t'))$ for all $t \geq t'$. This together with (6.12) imply $a_1(\varsigma) - x_{i1}(t) < \varepsilon$ for all $t \in [t', t_1 + \tau_D]$. In this time interval, we define a new vector $x'(t)$ by only replacing $x_{i1}(t)$ in $x(t)$ with $x'_{i1}(t) = a_1(\varsigma)$. Thus for all $t \in [t', t_1 + \tau_D]$,

$$\|x(t) - x'(t)\| = |x_{i1}(t) - a_1(\varsigma)| < \varepsilon,$$

and therefore,

$$|f_p^{i1}(x'(t))| - |f_p^{i1}(x(t))| \leq \|f_p(x'(t)) - f_p(x(t))\| \leq L\|x(t) - x'(t)\| < L\varepsilon. \tag{6.13}$$

From the definition of $d_\varsigma(\cdot)$, we know that for all $t \in [t', t_1 + \tau_D]$,

$$|f_p^{i1}(x'(t))| \geq d_\varsigma(\delta),$$

Combining the above inequality and (6.13), one has

$$|f_p^{i1}(x(t))| > d_\varsigma(\delta) - L\varepsilon.$$

Notice that

$$\varepsilon = \varphi(\delta) < \bar{\varphi}(\delta) \leq \frac{\tau_D d_\varsigma(\delta)}{\tau_D L + 1},$$

or what is the same,

$$\tau_D\big(d_\varsigma(\delta) - L\varepsilon\big) > \varepsilon > 0.$$

This implies that $f_p^{i1}(x(t))$ does not change sign in $[t_1, t_1 + \tau_D]$ and

$$|x_{i1}(t_1 + \tau_D) - x_{i1}(t_1)| = \int_{t_1}^{t_1 + \tau_D} |f_p^{i1}(x(\tau))| d\tau > \tau_D\big(d_\varsigma(\delta) - L\varepsilon\big) > \varepsilon,$$

which contradicts assumption (6.12). ∎

Proof of Theorem 6.6: ($\Longrightarrow$) To prove the contrapositive form, assume that $\mathcal{G}(t)$ is not uniformly quasi-strongly connected, that is, for all $T > 0$ there exists $t^* \geq 0$ such that the digraph $\mathcal{G}\big([t^*, t^* + T]\big)$ is not quasi-strongly connected. Then by its definition, in the digraph $\mathcal{G}\big([t^*, t^* + T]\big)$ there are two nodes i^* and j^* such that for any node k either i^* or j^* is not reachable from k. Let $\mathcal{V}_1$ be the subset of nodes from which i^* is reachable and $\mathcal{V}_2$ be the subset of nodes from which j^* is reachable. Obviously, $\mathcal{V}_1$ and $\mathcal{V}_2$ are disjoint. Moreover, for each node $i \in \mathcal{V}_1$ (resp. $\mathcal{V}_2$), the set of neighbors of agent i in the digraph $\mathcal{G}([t^*, t^* + T])$ is a subset of $\mathcal{V}_1$ (resp. $\mathcal{V}_2$). This implies that, for all $t \in [t^*, t^* + T]$, and for all $(i, j) \in \mathcal{V}_1 \times \mathcal{V}_2$, $\mathcal{N}_i(\sigma(t)) \subset \mathcal{V}_1$ and $\mathcal{N}_j(\sigma(t)) \subset \mathcal{V}_2$.

Choose any $z_1 \in \mathcal{X}$ and $z_2 \in \mathcal{X}$ such that $z_1 \neq z_2$, let $t_0 = t^*$, and pick any initial condition $x(t_0)$ such that

$$x_i(t_0) = \begin{cases} z_1, & \forall i \in \mathcal{V}_1, \\ z_2, & \forall i \in \mathcal{V}_2. \end{cases}$$

Then

$$x_i(t) = \begin{cases} z_1, & \forall i \in \mathcal{V}_1, \\ z_2, & \forall i \in \mathcal{V}_2, \end{cases} \quad \forall t \in [t_0, t_0 + T].$$

Let $c = \|x(t_0)\|$ and let ε be a positive scalar smaller than $\|z_1 - z_2\|/2$. We have thus found $\varepsilon > 0$ and $c > 0$ such that, for all $T > 0$, there exists t_0 such that

$$\big(\|x(t_0)\| \leq c\big) \text{ and } \big(x(t_0) \in \mathcal{X}^n\big),$$

but

$$\big(\exists t = t_0 + T\big) \, \|x(t)\|_\Omega > \epsilon.$$

This proves that the system (6.3) does not achieve state agreement in $\mathcal{X}$ uniformly.

($\Longleftarrow$) We show the sufficiency using ∞-norm for convenience. That is, $\forall \varepsilon > 0$, $\forall c > 0$, $\exists T^* > 0$ such that $\forall t_0 \geq 0$

$$x^0 \in \mathcal{X}_c^n \implies \big(\exists \zeta \in \mathcal{X}\big) \big(\forall t \geq t_0 + T^*\big) \big(\forall i\big) \|x_i(t) - \zeta\|_\infty \leq \varepsilon.$$

Let $\varepsilon > 0$, $c > 0$ be arbitrary. Then, there exist a class $\mathcal{KL}$ function $\gamma(\cdot, \cdot)$ and a class $\mathcal{K}$ function $\varphi(\cdot)$ satisfying the properties in Lemma 6.6 and Lemma 6.7, respectively. For any given $t_0 \geq 0$ and $x^0 \in \mathcal{X}_c^n$, consider the solution $x(t)$ of (6.3) with $x(t_0) = x^0$ and the following nonnegative vector function

$$V(x) = \begin{bmatrix} V_1(x) & \cdots & V_m(x) \end{bmatrix},$$

where $V_j(x) = a_j(x) - b_j(x)$, $j = 1, \ldots, m$. By Theorem 6.3, for any $t \geq t' \geq t_0$, $x_i(t) \in \mathcal{C}(x(t')) \subset \mathcal{X}_c$ for all i. Then it follows that $V_j(x(t))$, $j = 1, \ldots, m$, is non-increasing along the trajectory $x(t)$.

If $\mathcal{G}(t)$ is uniformly quasi-strongly connected, it follows that there is a $T' > 0$ such that for all t the union digraph $\mathcal{G}([t, t + T'])$ is quasi-strongly connected. Let $T = T' + 2\tau_D$, where τ_D is the dwell time. Then we are going to show there exists a class $\mathcal{K}$ function $\eta(\cdot)$ such that for any fixed $t' \geq t_0$

$$V_1(x(t' + \bar{T})) - V_1(x(t')) \leq -\eta\left(V_1(x(t'))\right), \qquad (6.14)$$

where $\bar{T} = 2nT$.

The proof relies on constructing a family of parameters $\varepsilon_1, \delta_2, \varepsilon_2, \ldots, \varepsilon_{n-1}, \delta_n, \varepsilon_n$ defined recursively as follows:

set $\varepsilon_n = \frac{V_1(x(t'))}{2}$;

for $k = n, \ldots, 2$

set $\delta_k = \gamma(\varepsilon_k, \bar{T})$;

set $\varepsilon_{k-1} = \varphi(\delta_k)$.

Define $\bar{\gamma}(\cdot) := \gamma(\cdot, \bar{T})$. Then ε_1 can be written as

$$\varepsilon_1 = \eta\left(V_1(x(t'))\right)$$

where $\eta(\cdot) := \varphi \circ \bar{\gamma} \circ \cdots \varphi \circ \bar{\gamma}(\frac{\cdot}{2})$. It is a class $\mathcal{K}$ function since $\bar{\gamma}(\cdot)$ and $\varphi(\cdot)$ are both class $\mathcal{K}$ functions. Since $\gamma(\cdot, \cdot)$ is class $\mathcal{KL}$ with the property $\gamma(\Delta, 0) = \Delta$ and $\bar{T} > 0$, it follows that $\delta_k < \varepsilon_k$. In addition, $\varepsilon_{k-1} < \delta_k$ because of $\varphi(\Delta) < \Delta$ for nonzero Δ. Thus,

$$0 < \varepsilon_1 < \delta_2 < \cdots < \delta_n < \varepsilon_n.$$

Recall that $\mathcal{H}_r(x) = \{y \in \mathcal{C}(x) : y_1 \leq a_1(x) - r\}$ for any small enough number r (in this proof, r is either ε_i or δ_i). In what follows, without causing confusion, we use $\mathcal{H}_r$ to denote $\mathcal{H}_r(x(t'))$ for simplicity.

We let

$$\tau_1 = t' + \tau_D$$

$$\tau_2 = t' + T + \tau_D$$

$$\vdots$$

$$\tau_{2n} = t' + (2n - 1)T + \tau_D.$$

See Figure 6.16 for an illustration. Notice that for each $k = 1, \ldots, 2n$, the digraph $\mathcal{G}([\tau_k, \tau_k + T'])$ is quasi-strongly connected. It therefore has a center by theorem 2.1, say c_k. Let $\mathcal{V}_1$ and $\mathcal{V}_1^*$ be a partition of the node set $\mathcal{V}$ such that

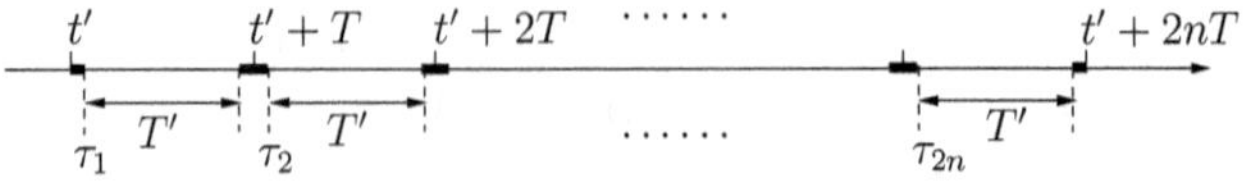

$$t'\qquad\quad t'+T\quad\ t'+2T\quad\cdots\cdots\qquad\qquad\qquad t'+2nT$$

$$\underbrace{\quad}_{\tau_1}\ T'\ \underbrace{\quad}_{\tau_2}\ T'\qquad\quad\cdots\cdots\qquad\quad\underbrace{\quad}_{\tau_{2n}}\ T'$$

Fig. 6.16. Time interval $[t', t' + \bar{T}]$.

$i \in \mathcal{V}_1$ if $x_i(t') \in \mathcal{H}_{\varepsilon_n}$ and $i \in \mathcal{V}_1^*$ otherwise. Thus, c_k is either in $\mathcal{V}_1$ or $\mathcal{V}_1^*$, so at least n elements in $\{c_1, \ldots, c_{2n}\}$ lie in either $\mathcal{V}_1$ or $\mathcal{V}_1^*$. Assume without loss of generality that they lie in $\mathcal{V}_1$, so there exist indices $1 \le k_1 < \cdots < k_n \le 2n$ such that $c_{k_i} \in \mathcal{V}_1$.

At time t', by its definition $\mathcal{H}_{\varepsilon_n}$ has at least one agent (see Fig. 6.17). Moreover, by Lemma 6.6, it follows that for all i

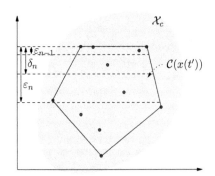

Fig. 6.17. Distribution of agents at time t'.

$$x_i(t') \in \mathcal{H}_{\varepsilon_n} \implies x_i(t) \in \mathcal{H}_{\delta_n}\ \forall t \in [t', t' + \bar{T}]. \qquad (6.15)$$

Recall that the digraph $\mathcal{G}([\tau_{k_1}, \tau_{k_1} + T'])$ has a center c_{k_1} in $\mathcal{V}_1$. Hence, there exists a pair $(i, j) \in \mathcal{V}_1^* \times \mathcal{V}_1$ such that j is a neighbor of i in this digraph since otherwise there is no arc from j to i for any $i \in \mathcal{V}_1^*$ and $j \in \mathcal{V}_1$, which contradicts the fact that the digraph has a center in $\mathcal{V}_1$. This further implies that there is a $\tau \in [\tau_{k_1}, \tau_{k_1} + T']$ such that $j \in \mathcal{N}_i(\tau)$. Since

$$\tau \in [\tau_{k_1}, \tau_{k_1} + T'] = [t' + (k_1 - 1)T + \tau_D, t' + k_1 T - \tau_D],$$

it follows that

$$[\tau - \tau_D, \tau + \tau_D] \subset [t' + (k_1 - 1)T, t' + k_1 T].$$

Since $\sigma(t) \in \mathcal{S}_{dwell}(\tau_D)$, there is an interval $[\bar{\tau}, \bar{\tau} + \tau_D]$, which contains τ and is a subinterval of $[t', t' + k_1 T]$, such that $j \in \mathcal{N}_i(t)$ for all $t \in [\bar{\tau}, \bar{\tau} + \tau_D]$. In addition, since $j \in \mathcal{V}_1$, or what is the same, $x_j(t') \in \mathcal{H}_{\varepsilon_n}$, from (6.15) we know that $x_j(t) \in \mathcal{H}_{\delta_n}$ for all $t \in [t', t' + \bar{T}]$ (and of course for all $t \in [\bar{\tau}, \bar{\tau} + \tau_D]$). Thus, by Lemma 6.7, there exists $t_1 \in [t', \bar{\tau} + \tau_D] \subseteq [t', t' + k_1 T]$ such that $x_i(t_1) \in \mathcal{H}_{\varepsilon_{n-1}}$.

On the one hand, we showed that the agent i not in $\mathcal{H}_{\varepsilon_n}$ at t' is in $\mathcal{H}_{\varepsilon_{n-1}}$ at t_1. On the other hand, the agents in $\mathcal{H}_{\varepsilon_n}$ at t' remain in $\mathcal{H}_{\delta_n}$ at t_1 from (6.15) and therefore remain in $\mathcal{H}_{\varepsilon_{n-1}}$ at t_1 because $\mathcal{H}_{\delta_n} \subset \mathcal{H}_{\varepsilon_{n-1}}$. Hence, at time t_1, $\mathcal{H}_{\varepsilon_{n-1}}(x(t'))$ has at least two agents.

Let $\mathcal{V}_2$ and $\mathcal{V}_2^*$ be a partition of the node set $\mathcal{V}$ such that $i \in \mathcal{V}_2$ if $x_i(t_1) \in \mathcal{H}_{\varepsilon_{n-1}}$ and $i \in \mathcal{V}_2^*$ otherwise. Note that by (6.15)

$$k \in \mathcal{V}_1 \Longrightarrow x_k(t') \in \mathcal{H}_{\varepsilon_n} \underset{(6.15)}{\Longrightarrow} x_k(t_1) \in \mathcal{H}_{\delta_n} \subset \mathcal{H}_{\varepsilon_{n-1}} \Longrightarrow k \in \mathcal{V}_2,$$

so $\mathcal{V}_1 \subset \mathcal{V}_2$. In particular c_{k_2}, the center node of $\mathcal{G}([\tau_{k_2}, \tau_{k_2} + T'])$, is in $\mathcal{V}_2$ because it is in $\mathcal{V}_1$. Then we can apply the same argument to conclude that there is a $t_2 \in [t_1, t' + k_2 T]$ such that $x_i(t_2) \in \mathcal{H}_{\varepsilon_{n-2}}$ and therefore, $\mathcal{H}_{\varepsilon_{n-2}}$ has at least three agents at t_2.

Repeating this argument $n - 1$ times leads to the result that there is a

$$t_{n-1} \in [t', t' + k_{n-1}T] \subset [t', t' + \bar{T}]$$

such that $\mathcal{H}_{\varepsilon_1}$ has n agents at t_{n-1}. Hence,

$$V_1(x(t_{n-1})) \leq V_1(x(t')) - \varepsilon_1 = V_1(x(t')) - \eta(V_1(x(t')))$$

and (6.14) follows.

Since we now know (6.14) holds, then we have

$$V_1\left(x(t_0 + k\bar{T})\right) \leq V_1\left(x(t_0)\right) - \eta\left(V_1(x(t_0))\right) - \cdots - \eta\left(V_1(x(t_0 + (k-1)\bar{T}))\right).$$

Notice that $x^0 \in \mathcal{X}_c^n(0)$ implies $V_1(x^0) \leq 2c$. In addition, considering the facts that $\eta(\cdot)$ is a class $\mathcal{K}$ function and that $V_1(x(t))$ is nonincreasing, one obtains

$$V_1(x(t_0 + k\bar{T})) \leq 2c - k\eta(V_1(x(t_0 + k\bar{T}))).$$

This means there is a $T_1^* = k\bar{T} > 0$ (k large enough) such that

$$V_1(x(t)) < 2\varepsilon \text{ for all } t \geq t_0 + T_1^*.$$

For each $j \in \{1, \ldots, m\}$, by the same argument, there is a $T_j^* > 0$ such that

$$V_j(x(t)) < 2\varepsilon \text{ for all } t \geq t_0 + T_j^*.$$

Let $T^* = \max\{T_j^*, \ j = 1, \ldots, m\}$. Thus

$$V_j(x(t)) < 2\varepsilon \text{ for all } t \geq t_0 + T^* \text{ and for all } j \in \{1, \ldots, m\}.$$

This in turn implies that there exists a $\zeta \in \mathcal{X}$ such that

$$\|x_i(t) - \zeta\|_\infty \leq \varepsilon \quad \text{for all } i \text{ and for all } t \geq t_0 + T^*.$$

We have thus proven that the system (6.3) achieves state agreement in $\mathcal{X}$ uniformly. ∎

In the remainder of this section we will present some examples to better illustrate the nature of our assumptions.

Concerning Assumption A1

We now present an example showing that Theorem 6.6 may fail to hold when the vector fields are just continuous instead of locally Lipschitz.

Consider three agents, 1, 2, and 3, with state space $\mathbb{R}$. There are three possible vector fields:

$$p = 1: \qquad\qquad p = 2: \qquad\qquad p = 3:$$

$$\left\{\begin{array}{l} \dot{x}_1 = g(x_3 - x_1) \\ \dot{x}_2 = 0 \\ \dot{x}_3 = 0 \end{array}\right\}, \quad \left\{\begin{array}{l} \dot{x}_1 = g(x_2 - x_1) \\ \dot{x}_2 = 0 \\ \dot{x}_3 = 0 \end{array}\right\}, \quad \left\{\begin{array}{l} \dot{x}_1 = 0 \\ \dot{x}_2 = g(x_1 - x_2) \\ \dot{x}_3 = 0 \end{array}\right\}$$

where

$$g(y) := \text{sign}(y) \cdot |y|^{\frac{1}{2}}, \ y \in \mathbb{R}.$$

The function g has the property that each solution of the differential equation $\dot{y} = g(y)$ reaches the origin (asymptotically stable equilibrium) in finite time.

For each $p \in \mathcal{P} = \{1, 2, 3\}$, the corresponding mode graphs are depicted in Fig. 6.18. Let $\mathcal{X} = \mathbb{R}$. Obviously, the function $g(\cdot)$ is only continuous (not locally Lipschitz on $\mathbb{R}$), so assumption A1 does not hold, but it can be easily checked that A2 holds. Let us set a switching signal $\sigma(t)$ to be periodic with period of

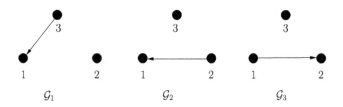

Fig. 6.18. Mode graphs $\mathcal{G}_p$, $p = 1, 2, 3$.

12 seconds, that is,

$$\sigma(t) = \begin{cases} 1, t \in [12k, 12k + 4), \\ 2, t \in [12k + 4, 12k + 8), \\ 3, t \in [12k + 8, 12k + 12), \end{cases} \quad k = 0, 1, \ldots.$$

Thus, assumption A3 holds.

For the switched nonlinear coupled cell system corresponding to the switching signal above, the interaction graph $\mathcal{G}(t)$ is uniformly quasi-strongly connected. To see that, simply let $T = 12$ and notice that for any $t > 0$,

$$\mathcal{G}([t, t + T]) = \mathcal{G}_1 \cup \mathcal{G}_2 \cup \mathcal{G}_3$$

is quasi-strongly connected. However, this switched nonlinear coupled cell system does not achieve state agreement uniformly as shown by a simulation in Fig. 6.19. Intuitively, for the period of $\sigma(t) = 1$, agent 1 moves toward agent 3 and the others remain stationary, whereas for the period of $\sigma(t) = 2$, agent 1 moves toward agent 2 and the others remain stationary. However, agent 1 reaches the location of agent 2 and stays there during this period. Then, when the system switches to $p = 3$, agent 2 starts to move toward agent 1, but since agents 1 and 2 are already collocated, agent 2 keeps stationary. Hence, only agent 1 moves

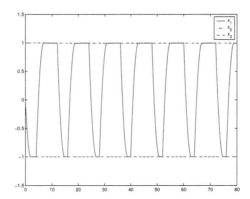

Fig. 6.19. Not tending to a common value.

forward and backward between the locations of agent 2 and 3 while the others are stationary.

Concerning Assumption A2

Our next example is concerned with the necessity of the strictness in assumption A2. This cannot be relaxed to just $f_p^i(x) \in \mathcal{T}(x_i, \mathcal{C}_p^i)$, as shown next.

Consider two agents, 1 and 2, with state space $\mathbb{R}$. There is only one vector field:

$$p = 1 : \left\{ \begin{array}{l} \dot{x}_1 = f_1^1(x_1, x_2) = 0 \\ \dot{x}_2 = f_1^2(x_1, x_2) = g(x_1 - x_2) \end{array} \right\}$$

where the smooth function $g : \mathbb{R} \to \mathbb{R}$ is given in Fig. 6.20.

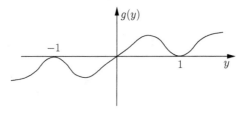

Fig. 6.20. Smooth function $g(y)$.

The nonlinear coupled cell system above has fixed coupling structure, that is, $\sigma(t) \equiv 1$. So assumption A3 is trivially satisfied. Let $\mathcal{X} = \mathbb{R}$. Assumption A1 holds, but A2 does not hold since

$$f_1^2(x_1, x_2) = g(x_1 - x_2) = 0 \notin \operatorname{ri}\big(\mathcal{T}(x_2, \mathcal{C}_1^2)\big)$$

when $x_1 = x_2 + 1$ by noticing that $\mathcal{C}_1^2 = \operatorname{co}\{x_1, x_2\}$ is the line segment joining x_1 and x_2. However, $f_1^1(x_1, x_2)$ and $f_1^2(x_1, x_2)$ are in $\mathcal{T}(x_1, \mathcal{C}_1^1)$ and $\mathcal{T}(x_2, \mathcal{C}_1^2)$ respectively for all $(x_1, x_2) \in \mathcal{X} \times \mathcal{X}$.

It can be easily checked that the interaction graph $\mathcal{G}(t)$ is fixed and is quasi-strongly connected, so it is uniformly quasi-strongly connected. But this non-linear coupled cell system does not globally achieve state agreement when, for example, initially $x_1(0) = x_2(0) + 1$.

However, if we choose $\mathcal{X} = [a, b]$, where a, b are real numbers satisfying $b - a < 1$, then assumptions A1, A2, and A3 hold. Thus, it follows that the system achieves state agreement in $\mathcal{X}^2$.

Concerning Assumption A3

Now we turn our attention to assumption A3. In order to guarantee convergence property, some regularity conditions on the switching signal $\sigma(\cdot)$ are needed. This is illustrated by the following very simple linear example.

Consider just two agents, 1 and 2, with state space $\mathbb{R}$. There are two possible vector fields:

$$p = 1: \left\{ \begin{array}{l} \dot{x}_1 = x_2 - x_1 \\ \dot{x}_2 = 0 \end{array} \right\}, \quad p = 2: \left\{ \begin{array}{l} \dot{x}_1 = 0 \\ \dot{x}_2 = 0 \end{array} \right\}$$

Thus agent 2 has no neighbor and never moves. For $p = 1$ agent 1 moves toward agent 2, whereas for $p = 2$ agent 1 has no neighbor and therefore doesn't move. Assumptions A1 and A2 hold for $\mathcal{X} = \mathbb{R}$. Let us define switching times τ_k by setting $\tau_0 = 0$ and defining the intervals $\delta_k = \tau_{k+1} - \tau_k$ as follows:

k	0	1	2	3	4	5	6	$\cdots$
δ_k	1	1	1/2	1	$1/2^2$	1	$1/2^3$	$\cdots$

Then we define $\sigma(t)$ to be the alternating sequence $1, 2, 1, 2, \ldots$ over the time intervals, respectively,

$$[\tau_0, \tau_1), [\tau_1, \tau_2), [\tau_2, \tau_3), [\tau_3, \tau_4), \ldots$$

This switching signal is piecewise constant and the interaction graph is uniformly quasi-strongly connected. However, if $x_1(0) \neq x_2(0)$, $x_1(t)$ does not converge to $x_2(t)$—agreement does not occur.

The example suggests that in order to achieve agreement, one needs to impose some restrictions on the admissible switching signals. One way to address this problem is to make sure that the switching signal has a dwell time, that is, there exists $\tau_D > 0$ such that

$$(\forall k)\ (\tau_{k+1} - \tau_k) \geq \tau_D.$$

This is precisely the assumption A3, and is ubiquitous in the switching control literatures (see, for example, [55, 57, 74]).

Concerning Nonautonomous Vector Fields

It may seem tempting to conjecture that our main result, Theorem 6.6, still holds for the nonautonomous nonlinear coupled cell system

$$\dot{x}(t) = f_{\sigma(t)}(t, x(t)),$$

which has nonautonomous vector fields. Again assume that A1, A2, and A3 hold where $f_p^i(x)$ is replaced by $f_p^i(t, x)$. Our next example shows that this is generally not true.

Consider again two agents, 1 and 2, with state space $\mathbb{R}$. Suppose there is only one nonautonomous vector field:

$$p = 1: \quad \left\{ \begin{array}{l} \dot{x}_1 = e^{\alpha t}(x_2 - x_1) \\ \dot{x}_2 = e^{\beta t}(x_1 - x_2) \end{array} \right\}, \quad \text{where } \alpha, \beta < 0.$$

Thus, assumptions A1 and A2 hold for $\mathcal{X} = \mathbb{R}$. Here $\sigma(t) \equiv 1$ and so A3 is trivially satisfied. Moreover, the interaction graph $\mathcal{G}(t) = \mathcal{G}_1$ is uniformly quasi-strongly connected as required in Theorem 6.6. However, the system does not achieve state agreement, as shown in the simulation in Fig. 6.21 where

$$\alpha = -3, \ \beta = -2, \quad x_1(0) = 2, \ x_2(0) = 0.$$

Nevertheless, if we modify the sub-tangentiality assumption A2 on the nonautonomous vector fields by saying that for each $i \in \mathcal{V}$ and each $p \in \mathcal{P}$,

$$f_p^i(t, x) \in \mathrm{ri}(\mathcal{T}(x_i, \mathcal{C}_p^i)), \ \forall t \quad \text{and} \quad \inf_t f_p^i(t, x) \in \mathrm{ri}(\mathcal{T}(x_i, \mathcal{C}_p^i)),$$

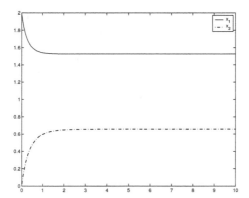

Fig. 6.21. Not tending to a common value.

then Theorem 6.6 may still hold. This assumption rules out situations like the one illustrated in this example. Here,

$$\inf_t f_1^1(x) = \inf_t e^{\alpha t}(x_2 - x_1) = 0 \text{ and } \inf_t f_1^2(x) = \inf_t e^{\beta t}(x_1 - x_2) = 0.$$

6.6 Synchronization of Coupled Oscillators

The Kuramoto model describes the dynamics of a set of n phase oscillators θ_i with natural frequencies ω_i. More details can be found in [59, 127]. The time evolution of the i-th oscillator is given by

$$\dot{\theta}_i = \omega_i + k_i \sum_{j \in \mathcal{N}_i(t)} \sin(\theta_j - \theta_i),$$

where $k_i > 0$ is the coupling strength and $\mathcal{N}_i(t)$ is the set of neighbors of oscillator i at time t. The interaction structure can be general so far, that is, $\mathcal{N}_i(t)$ can be an arbitrary set of other nodes and can be dynamic.

The neighbor sets $\mathcal{N}_i(t)$ define an interaction graph $\mathcal{G}(t)$ and the switched nonlinear coupled cell system

$$\dot{\theta}(t) = f_{\sigma(t)}(\theta(t))$$

where $\theta = (\theta_1, \cdots, \theta_n)$ and $\sigma(t)$ is a suitable switching signal. For identical coupled oscillators (i.e., $\omega_i = \omega, \forall i$), the transformation $x_i = \theta_i - \omega t$ yields

$$\dot{x}_i = k_i \sum_{j \in \mathcal{N}_i(t)} \sin(x_j - x_i), \quad i = 1, \ldots, n. \tag{6.16}$$

Let a, b be any real numbers such that $0 \leq b - a < \pi$, and define $\mathcal{X} = [a, b]$. It is easily seen that assumptions A1 and A2 are satisfied. Suppose $\sigma(t)$ here is regular enough satisfying A3. Then from Theorem 6.6 it follows that, if and only if, the interaction graph $\mathcal{G}(t)$ is uniformly quasi-strongly connected, the switched nonlinear coupled cell system (6.16) achieves state agreement in $\mathcal{X}$ uniformly. This implies that there exists $\zeta \in \mathbb{R}$ such that

$$\theta_i(t) \rightarrow \zeta + \omega t, \quad \dot{\theta}_i(t) \rightarrow \omega,$$

and the oscillators synchronize. This is an extension of Theorem 1 in [59], which assumes the interaction graph is bidirectional and static, and the initial state $\theta_i(0) \in \left(-\frac{\pi}{2}, \frac{\pi}{2}\right)$ for all i.

As an example, three Kuramoto oscillators with interaction structure are simulated. The initial conditions are $\theta_1 = 0$, $\theta_2 = 1$, $\theta_3 = -1$. The natural frequency ω_i equals 1, and the coupling strength k_i is set to 1 for all i. The interaction structure switches among three possible interaction structures periodically, shown in Fig. 6.22. It can be checked that $\mathcal{G}(t)$ is uniformly quasi-strongly connected. So

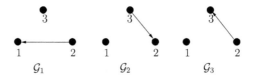

Fig. 6.22. Interaction graphs $\mathcal{G}_p$, $p = 1, 2, 3$.

these three oscillators achieve asymptotical synchronization as we conclude by our main theorem. Fig. 6.23 shows the plots of $\sin(\theta_i)$, $i = 1, 2, 3$ and of the switching signal $\sigma(t)$. Synchronization is evident.

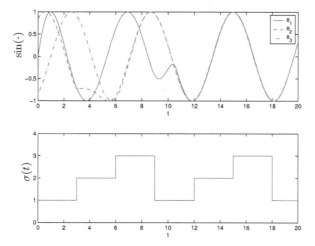

Fig. 6.23. Synchronization of three oscillators.

6.7 Biochemical Reaction Networks

A biochemical reaction network is a finite set of reactions among a finite set of species. Consider, for example, two reversible reactions among three compounds C_1, C_2, and C_3, in which C_1 is transformed into C_2, C_2 is transformed into C_3, and vice versa:

$$C_1 \underset{k_2}{\overset{k_1}{\rightleftharpoons}} C_2 \underset{k_4}{\overset{k_3}{\rightleftharpoons}} C_3$$

The constants $k_1 > 0$, $k_2 > 0$ are the forward and reverse rate constants of the reaction $C_1 \rightleftharpoons C_2$; similarly for $k_3 > 0$, $k_4 > 0$. Denote the concentrations of C_1, C_2, and C_3, respectively, by x_1, x_2, and x_3. Only nonnegative concentrations are physically possible. Such a reaction network gives rise to a dynamical system, which describes how the state of the network changes over time.

Suppose the dynamics of both reactions are dictated by the mass action principle. This leads to the model

$$\begin{aligned}
\dot{x}_1 &= -k_1 x_1^\alpha + k_2 x_2^\alpha \\
\dot{x}_2 &= k_1 x_1^\alpha - k_2 x_2^\alpha - k_3 x_2^\alpha + k_4 x_3^\alpha \\
\dot{x}_3 &= k_3 x_2^\alpha - k_4 x_3^\alpha
\end{aligned} \qquad (6.17)$$

where $\alpha \geq 1$ is an integer. For more on modeling and analysis of biochemical reaction networks, we refer to $[1, 60, 72, 143]$.

The vector field above obviously does not satisfy assumption A2. However, as we will see later, by a simple coordinate transformation, it will satisfy the assumption and we can then apply our main result directly to analyze the biochemical reaction network. For this purpose, we rewrite the above model as

$$\dot{x}_1 = g_1(x_1, x_2)(k_2^{\frac{1}{\alpha}} x_2 - k_1^{\frac{1}{\alpha}} x_1)$$
$$\dot{x}_2 = g_2(x_1, x_2)(k_1^{\frac{1}{\alpha}} x_1 - k_2^{\frac{1}{\alpha}} x_2) + g_3(x_2, x_3)(k_4^{\frac{1}{\alpha}} x_3 - k_3^{\frac{1}{\alpha}} x_2)$$
$$\dot{x}_3 = g_4(x_2, x_3)(k_3^{\frac{1}{\alpha}} x_2 - k_4^{\frac{1}{\alpha}} x_3)$$

where $g_1(x_1, x_2)$, $g_2(x_1, x_2)$, $g_3(x_2, x_3)$, and $g_4(x_2, x_3)$ are suitable terms. It can be easily verified that $g_1(x_1, x_2) \geq 0$ and $g_1(x_1, x_2) = 0$ only for the case $x_1 = x_2 = 0$ since x_1, x_2 are nonnegative. The same observations hold for $g_2(x_1, x_2)$, $g_3(x_2, x_3)$, and $g_4(x_2, x_3)$. It follows that when the state $x = (x_1, x_2, x_3)$ is at any point of the set

$$\Xi = \left\{ \zeta \in \mathbb{R}^3 : \zeta = a \begin{pmatrix} (\frac{k_2}{k_1})^{\frac{1}{\alpha}} \\ 1 \\ (\frac{k_3}{k_4})^{\frac{1}{\alpha}} \end{pmatrix}, \ a \geq 0 \right\},$$

the forward and the backward reaction are balanced at this point. A net change of concentration is not measurable from a macroscopic point of view, but in a microscopic view the reactions are still going on. This special state is called chemical equilibrium. Apply the coordinate transformation

$$y_1 = \left(\frac{k_1}{k_2}\right)^{\frac{1}{\alpha}} x_1, \quad y_2 = x_2, \quad y_3 = \left(\frac{k_4}{k_3}\right)^{\frac{1}{\alpha}} x_3,$$

and let
$$h_1(y_1, y_2) = k_1^{\frac{1}{\alpha}} g_1(x_1, x_2), \quad h_2(y_1, y_2) = k_2^{\frac{1}{\alpha}} g_2(x_1, x_2)$$
$$h_3(y_2, y_3) = k_3^{\frac{1}{\alpha}} g_3(x_2, x_3), \quad h_4(y_2, y_3) = k_4^{\frac{1}{\alpha}} g_4(x_2, x_3).$$

We then obtain

$$\dot{y}_1 = h_1(y_1, y_2)(y_2 - y_1)$$
$$\dot{y}_2 = h_2(y_1, y_2)(y_1 - y_2) + h_3(y_2, y_3)(y_3 - y_2) \qquad (6.18)$$
$$\dot{y}_3 = h_4(y_2, y_3)(y_2 - y_3)$$

where $h_1(y_1, y_2)$, $h_2(y_1, y_2)$, $h_3(y_2, y_3)$, and $h_4(y_2, y_3)$ are suitable terms; for example

$$h_1(y_1, y_2) = \left(\frac{k_1^{1/\alpha} k_2}{k_2^{1/\alpha}} \right) \frac{y_2^\alpha - y_1^\alpha}{y_2 - y_1}.$$

It can be easily verified that $h_1(y_1, y_2) \geq 0$ and $h_1(y_1, y_2) = 0$ if and only if $y_1 = y_2 = 0$. The same observations hold for $h_2(y_1, y_2)$, $h_3(y_2, y_3)$, and $h_4(y_2, y_3)$. It thus follows that the set $\Omega = \left\{ y \in \mathbb{R}_+^3 : y_1 = y_2 = y_3 \right\}$ is an equilibria set. Physically, when $y \in \Omega$, the reaction network is at a chemical equilibrium.

Consider now the interaction graph associated with (6.18). Physically, each node represents a compound and each arc connecting two nodes represents a reaction between two compounds. This digraph is quasi-strongly connected (actually, it is strongly connected). Since there is no switching in the system (i.e., $\sigma(t)$ is constant), assumption A3 is obviously satisfied and the interaction graph is uniformly quasi-strongly connected. In addition, it can be easily checked that, for $\mathcal{X} = [0, \infty)$, the vector field in the above system satisfies assumptions A1 and A2. Hence, Theorem 6.6 can be applied to conclude that system (6.18) achieves state agreement in $\mathcal{X}^3$. This implies that the set Ξ is asymptotically stable for system (6.17) with region of stability $\mathbb{R}_+^3$. This result coincides with the analysis using Theorem 5.2 in [1]. Our analysis can be extended to more complicated biochemical reaction networks containing a set of compounds and a set of reversible reactions. Their asymptotic state agreement property is captured by the interaction graph.

6.8 Water Tank Networks

Water tank networks are interesting to study from the point of view of coupled cell systems. Consider a tank of water and suppose the water level is x (see Fig. 6.24). Then the pressure at the outlet is proportional to $\sqrt{x}$, so the flow rate out is proportional to $\sqrt{x}$ (Toricelli's Law), which gives rise to the following continuous-time model

$$\dot{x} = -a\sqrt{x},$$

where $a > 0$.

Now we consider two identical coupled tanks (see Fig. 6.25). The flow rate of tank 1 is a function of $x_2 - x_1$. Thus

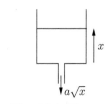

Fig. 6.24. Water tank.

Fig. 6.25. Two identical coupled tanks.

$$\dot{x}_1 = ag(x_2 - x_1)$$
$$\dot{x}_2 = ag(x_1 - x_2)$$

where $g(y) = \text{sign}(y) \cdot \sqrt{|y|}$. Refer x_2 to x_1 by defining the difference $d = x_2 - x_1$. Then the model for d is

$$\dot{d} = f(d),$$

where

$$f(d) = \begin{cases} 2\sqrt{-d}, & d \le 0 \\ -2\sqrt{d}, & d > 0. \end{cases}$$

So f is continuous but not Lipschitz (see Fig. 6.26).

It would be nice to extend this to more tanks (Fig. 6.27) and then to time-varying coupling (Fig. 6.28).

Now consider hundreds of water tanks connected through a network of pipes (see Fig. 6.29). Thus we have a general coupled nonlinear system model as follows

$$\dot{x}_1 = \sum_{j \in \mathcal{N}_1} a_{1j} g(x_j - x_1)$$

$$\vdots$$

$$\dot{x}_n = \sum_{j \in \mathcal{N}_n} a_{nj} g(x_j - x_n),$$

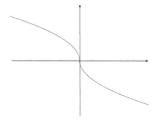

Fig. 6.26. Continuous but not Lipschitz.

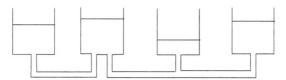

Fig. 6.27. Water tank networks with fixed coupling.

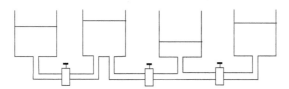

Fig. 6.28. Water tank networks with time-varying coupling.

where $a_{ij} > 0$ and $\mathcal{N}_i$ represents the set of these water tanks connected to the ith tank. It can be checked that each individual vector field is continuous

hundreds of water tanks

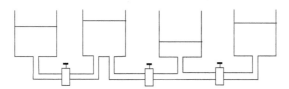

Fig. 6.29. Hundreds of water tanks.

(satisfying assumption A1′) though it is not locally Lipschitz and that it satisfies assumption A2′. The network topology naturally leads to an interaction graph which is completely the same as the one by our definition. By Theorem 6.2, the water level of the network of tanks will eventually be equalized if and only if the interaction graph is quasi-strongly connected (indeed it is equivalent to connected in the context of undirected graph since the graph is bidirectional). This is consistent to our experience.

Now suppose there are valves in some or all the pipes. And suppose at any time each valve can be either fully open or fully closed, and the valves may switch from one to the other. Thus it gives rise to a switched system model as in (6.3). Unfortunately, it does not satisfy local Lipschitz assumption and so Theorem 6.6 is not applicable. Indeed we once present a counterexample using this function $g(\cdot)$ to show that the failure of local Lipschitz continuity can cause the failure of Theorem 6.6. That example has a non-bidirectional interaction structure. However, by our experience, in the example of water tank network with switching, the water level will be eventually equalized if and only if the interaction graph is uniformly quasi-strongly connected. It is suggestive to conjecture that the theorem may still hold for bidirectional coupling structure with only continuity assumption, which we have not studied in depth.

6.9 Notes and Discussion

For nonlinear coupled cell systems, most existing work has dealt with static graphs having a particular topology, such as rings [16, 106], cyclic digraphs [112], and fully-connected graphs [46, 47, 124], or with static graphs having an unspecified topology but a certain connectedness, e.g., coupled cell networks [126] and coupled oscillators [59, 146]. A static graph simplifies the problem and allows one to focus on the difficulties caused by the nonlinear dynamics of the nodes. However, the more interesting situation is when the interaction graph is time-varying. Subsystems become disconnected from each other and may again become connected to each other for various natural or technological reasons.

From theoretic point of view, new tools are required for studying nonlinear coupled cell systems. Passivity as an analysis tool for nonlinear coupled systems is considered in [7]. The agreement protocol is given by

$$\dot{x}_i = H_i \left\{ -\sum_{j \in \mathcal{N}_i} \psi_j(x_i - x_j) \right\}, \quad i = 1, \ldots, n$$

where $\mathcal{N}_i$ is the neighbor set, the function $\psi_j(z_j)$ is selected to be $\nabla P_j(z_j)$—the gradient vector of a positive definite, radially unbounded function $P_j : \mathbb{R}^n \to \mathbb{R}$ that satisfies

$$z_j^T \nabla P_j(z_j) > 0, \quad \forall z_j \neq 0. \qquad \text{(sector nonlinearity)}$$

$H_i\{u_i\}$ denotes the output of a static or dynamic block (Fig. 6.30) satisfying strictly passive assumption. By passivity theory, whether all variables are syn-

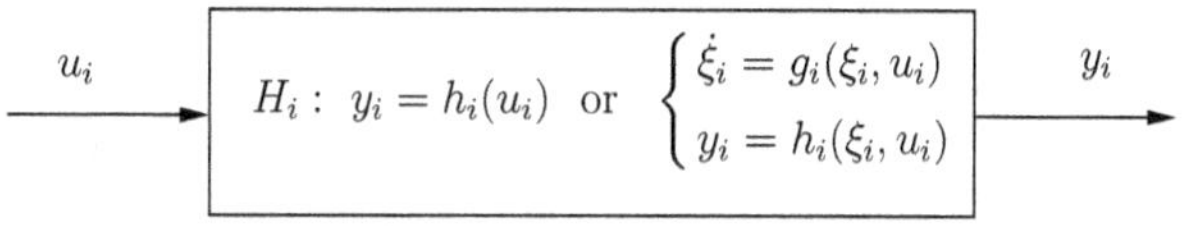

Fig. 6.30. Block H_i.

chronizable in the limit is then concluded. The result encompasses the one of [102] where the system does not have the block $H_i\{\cdot\}$. Because $H_i\{\cdot\}$ can be dynamic, the result is applicable to plants with higher-order dynamics than an integrator.

In discrete-time, remarkable results are obtained in [96] and [6]. In [96], a central assumption is imposed. That is, each cell's vector field satisfies a convex assumption, namely,

$$x_i(k+1) \in e_i(k)(x(k)), \quad i = 1, \ldots, n$$

where $e_i(k)(x(k))$ is a compact set contained in the convex hull of cell i's current state and its neighbors' states. This assumption then leads to the development of a set-valued Lyapunov function method, with the convex hull of the individual cells' states playing the role of a nonincreasing set-valued Lyapunov function in convergence analysis. In [6], a new setting is proposed to allow the presence of arbitrary bounded delays in the dynamic systems. The system in [6] is considered to have delays smaller than a given integer $h > 0$. In consequence, the complete state variable of the system is

$$(x_1(k), x_1(k-1), \ldots, x_1(k-h+1), \ldots, x_n(k), \ldots, x_n(k-h+1)) \in \mathbb{R}^{hn},$$

denoted by $\tilde{x}(k)$. Then a directed multigraph is considered to model linking some past and/or present values $x_i(k-m)$ of the states of an agent i to another agent j. Fig. 6.31 provides an example. For the graph represented in, agent 1 receives

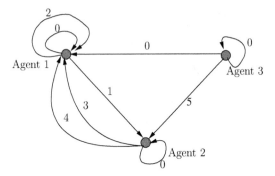

Fig. 6.31. Directed multigraph.

two delayed values of agent 2 where the number on arcs means the time delay (i.e., $x_2(k-3)$ and $x_2(k-4)$ are received by agent 1 at time k). Thus, the state update rule is written as

$$x_i(k+1) \in e_i(k)(\tilde{x}(k)), \quad i = 1, \ldots, n.$$

Here, the continuous compact set-valued map $e_i(k)(\tilde{x}(k))$ is contained in a set that is not necessary a convex hull but satisfies certain properties. These properties are used to forbid increase of the natural set-valued Lyapunov function along the trajectory and also to induce strict decrease of the set-valued Lyapunov function.

Part IV

Distributed Control of Autonomous Agents

7

Collective Point-Mass Robots

The problem of coordinated control of a group of autonomous wheeled vehicles is of recent interest in control and robotics. For this problem, the vehicles in the group are indeed dynamically decoupled, meaning that the motion of one vehicle does not affect another, but they are coupled through information exchange in order to achieve some desired cooperative tasks. So the problem of coordinated and cooperative control of multiple vehicles is within the more general subject— coupled cell systems. One interesting aspect of this subject involves the structure of the information flow among the vehicles, which plays a very important role in solving some coordination problems such as formation control and rendezvous.

In this chapter, the simple point-mass robots are considered to clarify the design philosophy without getting bogged down in vehicle dynamics.

7.1 State Model, Local Measurement and Control

The point-mass robot acts in continuous time, moves in the x, y-plane, and is linear and time-invariant (LTI):

Assume the velocities are controllable:

$$\begin{cases} \dot{x}_i = v_i \\ \dot{y}_i = w_i. \end{cases} \tag{7.1}$$

Write in vector form:

$$z_i = \begin{bmatrix} x_i \\ y_i \end{bmatrix}, \quad u_i = \begin{bmatrix} v_i \\ w_i \end{bmatrix}, \quad \dot{z}_i = u_i,$$

or equivalently, in complex variable form:

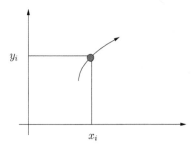

Fig. 7.1. Point mass robot in x, y-plane.

$$z_i = x_i + jy_i, \quad u_i = v_i + jw_i, \quad \dot{z}_i = u_i.$$

The point-mass model above captures a class of realistic mobile robots such as fully actuated or omni-directional robots [52]. Consider a group of n robots in a planar world with a global coordinate system, Σ. Suppose each robot is self-driven and controlled by the velocities v^i and w^i. As shown in Fig. 7.2,

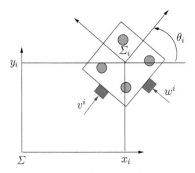

Fig. 7.2. Fully actuated mobile robot.

it can be implemented with four low-friction, omni-directional casters and two mounted high-performance ducted fans each capable of producing continuous thrust. We now construct a local frame Σ_i fixed on the robot, whose x and y-axis are consistent with the directions of v^i and w^i, respectively. Robot i is at location z_i with respect to Σ and it has orientation θ_i (θ_i is constant in the setup). Let $u^i = v^i + jw^i$. Then in complex variable form,

$$\dot{z}_i = e^{j\theta_i} u^i.$$

Suppose each robot has onboard an omni-directional camera such that, for example, the x-axis of local frame Σ_i points toward the center of the camera's field of view. Thus, if robot j is in the field of view of robot i, the local measurement by robot i is the relative position with respect to Σ_i, namely,

$$e^{-j\theta_i}(z_j - z_i).$$

Shown in Fig. 7.3 are two robots, a global frame Σ, and two local frames, Σ_1, Σ_2, rotated by θ_1, θ_2, respectively. Robot 1 sees robot 2 at the location $e^{-j\theta_1}(z_2 - z_1)$ with respect to Σ_1.

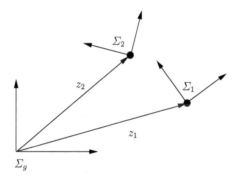

Fig. 7.3. Global vs local coordinate.

Consider the control strategy that robot 1 heads for robot 2 with the speed proportional to the distance between them. That is, robot 1 applies the velocity input

$$u^1 = ke^{-j\theta_1}(z_2 - z_1).$$

So in the global coordinate system Σ, it leads to

$$\dot{z}_1 = e^{j\theta_1} u^1 = k(z_2 - z_1),$$

which is consistent to the one obtained by directly applying $u_1 = k(z_2 - z_1)$ in eq. (7.1). That means, the control law $u_1 = k(z_2 - z_1)$ defined in the global frame is locally implementable by robot 1 using only local measurement. Therefore, in what follows, we will just consider the point-mass model (7.1) and assume each robot measures the relative position in a global coordinate system.

7.2 Information Flow Graph

For a group of n robots in the plane, they are indeed dynamically decoupled. That is, the motion of one robot does not directly affect any of the other robots if we do not take collision into account. However, they are coupled through the information flow in the group in order to achieve desired group behavior such as formation.

Now we define the information flow graph for the group of n robot. We refer to individual robots as nodes and information flows as links.

Definition 7.1. *The* information flow graph $\mathcal{G}$ *consists of*

- *a node set* $\mathcal{V} = \{1, 2 \ldots, n\}$, *each node i corresponding to robot i;*
- *a set $\mathcal{E}$ of arcs: The arc from node j to node i is one of its arcs just in case robot i can access the information from robot j in some way.*

By this definition, information flows in the direction of the arcs in $\mathcal{G}$. Let $\mathcal{N}_i$ denote the set of labels of those robots whose information flows to robot i.

For different sensing and/or communication schemes, the information flow graph for the group of n robots might be fixed, or dynamically changing depending on time, the state/output of the robots, and/or possibly some external signals. In what follows, if the information flow graph only depends on time, we denote $\mathcal{G}(t)$; if it only depends on the state z of the n robots, we denote $\mathcal{G}(z)$; and if it depends on both, we denote $\mathcal{G}(t, z)$. It should be pointed out that the information flow graph is called *proximity graph* in some literature when it only depends on the position of the robots. Surely, different type of information flow graph may produce totally different system behaviors.

For example, consider a group of six robots that want to agree upon certain quantities of interest, say the speed. Suppose only one-way point to point communication is allowed. In other words, every robot is allowed to either send the data to or receive the data from another one at any time. So it is impossible to have a fixed communication link so that the information flow graph is connected in some sense. Then an alternative communication scheme is proposed: Periodically switch between two communication links that are shown in Fig. 7.4, where

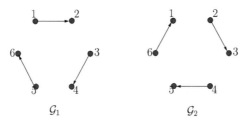

Fig. 7.4. Communication links.

the direction in the graph is the direction of data transferring. This would lead to a time-dependent information flow graph $\mathcal{G}(t)$.

Suppose now that n robots carry omnidirectional cameras of identical range r. Fig. 7.5 shows the disk-like field of view of robot 2 and the information flow graph at that time. Thus, the information flow graph is solely determined by the position of the n robots and therefore it is a state-dependent information flow graph.

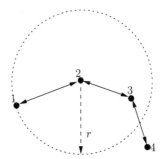

Fig. 7.5. Disk-like field of view.

7.3 Formation Evolution via Cyclic Pursuit

The French mathematician Pierre Bouguer [1698 - 1758] studied pursuit curves in 1732. As an example, in Fig. 7.6 point A moves at unit speed and point P pursues A at unit speed. Pursuit means that a pursuer must always head directly

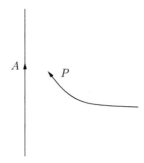

Fig. 7.6. Pursuit curve.

toward the pursued and the pursuer's speed must be proportional to or match that of the pursued:

$$\dot{z}_P = \frac{z_A - z_P}{\|z_A - z_P\|},$$

where z_p and z_A are the position of P and A, respectively. The trajectory of P is a *pursuit curve*. We extend this idea to robots.

Consider n labelled robots and consider the *cyclic pursuit* strategy. In other words, robot 1 pursues robot 2, robot 2 pursues robot 3, ..., and finally, robot n pursues robot 1. Let $z_i \in \mathbb{C}$ represent the position of robot i in the plane. Then the following closed-loop system describes the dynamics of n cyclic pursuing robots:

$$\begin{cases} \dot{z}_1 = u_1 = z_2 - z_1 \\ \quad\vdots \\ \dot{z}_n = u_n = z_1 - z_n \end{cases} \qquad (7.2)$$

As we proved in Theorem 5.1, the system achieves state agreement globally and each robot converges to the centroid of their initial locations. For the cyclic pursuit strategy, control is distributed (identical local strategies) with minimal sensor requirements (n information flow links for n robots). However, there is an emergent behavior: convergence to a common point.

Here are two simulations to illustrate the behavior. Fig. 7.7 shows the pursuit curves of six robots that are initially ordered and numbered clockwise, and the pursuit curves shown in Fig. 7.8 are resulting from ten robots with random initial arrangement. We observe that the trajectories sometimes intersect and sometimes do not, depending on initial configuration.

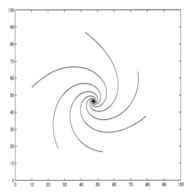

Fig. 7.7. Pursuit curves of six robots initially arranged clockwise.

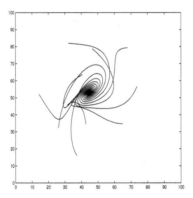

Fig. 7.8. Pursuit curves of ten robots initially arranged randomly.

So far, robots are points. So collisions are rare! Still, it is interesting to study the time evolution of formations. The case $n = 2$ is trivial—the two robots move in a straight line toward each other. If four robots are initially in a square, they remain so. More generally, if the robots are initially arranged in a convex polygon formation, they remain so. We intend now to extend this result.

We say that n robots are in a *counterclockwise star formation* if they are arranged like in Fig. 7.9: That is, all $r_i > 0$, all $\alpha_i > 0$, and $\sum_{i=1}^n \alpha_i = 2\pi$. Here are some examples in Fig. 7.10. Obviously, a convex polygon is a special case

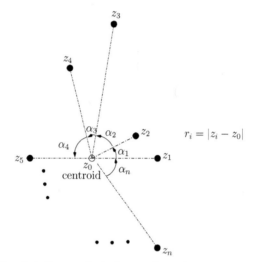

Fig. 7.9. Counterclockwise star formation.

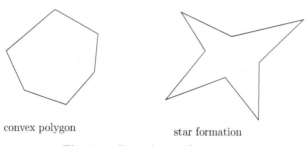

convex polygon   star formation

Fig. 7.10. Example: star formations.

of a star formation. A *clockwise star formation* is defined in a similar way. We consider only counterclockwise star formations since clockwise star formations can be treated analogously.

Theorem 7.1. *Suppose that n (> 2) distinct robots initially are arranged in a counterclockwise star formation. Under cyclic pursuit the robots remain in a counterclockwise star formation. (In particular, they never collide.)*

Let us prove the theorem when $n = 3$, which is not too hard. Then the theorem says: If the three robots initially form a triangle, then they always do so, that is, the triangle never collapses to a line or point.

Proof: Assume the initial configuration of three robots is a triangle as shown in Fig. 7.11. Define $F := \mathrm{Im}\left[(\overline{z_1 - z_3})(z_2 - z_3)\right]$ ($\mathrm{Im}[\cdot]$ denotes the imaginary part

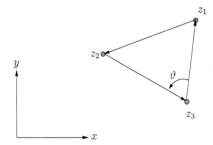

Fig. 7.11. Initial configuration.

and $\overline{z_1 - z_3}$ denotes the conjugate of $z_1 - z_3$) and bring in polar coordinates:

$$z_1 - z_3 = r_3^1 e^{j\theta_3^1}, \quad z_2 - z_3 = r_3^2 e^{j\theta_3^2}.$$

Then

$$F = \mathrm{Im}\left[r_3^1 r_3^2 e^{j(\theta_3^2 - \theta_3^1)}\right] = \mathrm{Im}\left[r_3^1 r_3^2 e^{j\vartheta}\right] = r_3^1 r_3^2 \sin(\vartheta).$$

Since the robots move, ϑ and F are functions of t. By assumption, $0 < \vartheta(0) < \pi$, and so $F(0) > 0$. Taking the derivative of F along the trajectory of system (7.2), we have

$$\dot{F} = \tfrac{d}{dt}\mathrm{Im}\left[(\overline{z_1 - z_3})(z_2 - z_3)\right]$$

$$= \mathrm{Im}\left[(\overline{\dot{z}_1 - \dot{z}_3})(z_2 - z_3)\right] + \mathrm{Im}\left[(\overline{z_1 - z_3})(\dot{z}_2 - \dot{z}_3)\right]$$

$$= \mathrm{Im}\left[(\overline{(z_2 - z_1) - (z_1 - z_3)})(z_2 - z_3)\right] + \mathrm{Im}\left[(\overline{z_1 - z_3})((z_3 - z_2) - (z_1 - z_3))\right]$$

$$= \mathrm{Im}\left[(\overline{z_2 - z_1})(z_2 - z_3)\right] - F - F - 0$$

$$= \mathrm{Im}\left[(\overline{(z_2 - z_3) - (z_1 - z_3)})(z_2 - z_3)\right] - 2F$$

$$= -3F.$$

Thus $F(t) = e^{-3t}F(0) > 0$ for all $t \geq 0$, and so $0 < \vartheta(t) < \pi$ for all $t \geq 0$ that means the triangle never collapse to a line. ∎

The proof of the theorem for $n > 3$ requires some machinery. We begin with a tool for studying angles. Consider the setup in Fig. 7.12.

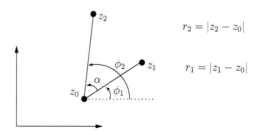

$$r_2 = |z_2 - z_0|$$

$$r_1 = |z_1 - z_0|$$

Fig. 7.12. Setup for Lemma 7.1.

Lemma 7.1. *Define* $F := \mathrm{Im}\left[(\overline{z_1 - z_0})(z_2 - z_0)\right]$. *Then*

(a) $0 < \alpha < \pi$, $r_1 > 0$, and $r_2 > 0$ if and only if $F > 0$,

(b) $\pi < \alpha < 2\pi$, $r_1 > 0$, and $r_2 > 0$ if and only if $F < 0$,

(c) *the points are collinear if and only if* $F = 0$.

Proof: Introduce polar coordinates:

$$z_1 - z_0 = r_1 e^{j\phi_1}, \quad z_2 - z_0 = r_2 e^{j\phi_2}.$$

Then

$$F = \mathrm{Im}\left[(\overline{z_1 - z_0})(z_2 - z_0)\right] = \mathrm{Im}\left[r_1 r_2 e^{j(\phi_2 - \phi_1)}\right] = r_1 r_2 \sin(\alpha).$$

Thus, $r_1 > 0$, $r_3 > 0$, and $0 < \alpha < \pi$ if and only if $F > 0$; and $r_1 > 0$, $r_3 > 0$ and $\pi < \alpha < 2\pi$ if and only if $F < 0$. Also, $F = 0$ if and only if $\alpha = 0$, $\alpha = \pi$, $r_1 = 0$, or $r_3 = 0$, i.e., the points are collinear. ∎

Lemma 7.2. *Suppose that n (> 2) points $z_1, \ldots, z_n$ form a counterclockwise star formation. Then $\alpha_i < \pi$, $i = 1, \ldots, n$.*

Proof: Suppose, e.g., $\alpha_1 \geq \pi$. Fix a coordinate system centered at z_0 with the positive real axis given by the ray from z_0 passing through z_1 (see Fig. 7.13). Then we have $\text{Im}[z_1] = 0$, $\text{Im}[z_2] \leq 0$, and $\text{Im}[z_k] < 0$ for $k = 3, \ldots, n$. Hence,

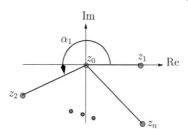

Fig. 7.13. Impossible setup.

$\text{Im}[z_0] = \sum_{i=1}^{n} \frac{\text{Im}[z_i]}{n} < 0$, a contradiction.   ∎

Lemma 7.3. *If $z_1, \ldots, z_n$ are collinear at some time, they are collinear for all time.*

Proof: Suppose the points are collinear at t_0. Re-orient the coordinate system so that the points at t_0 lie on the real axis $\mathbb{R}$. Considering the system (7.2), $\mathbb{R}^n$ is an invariant subspace. Hence, for any $i = 1, \ldots, n$, $z_i(t) \in \mathbb{R}$ for all t, implying the points are collinear for all time.   ∎

Proof of Theorem 7.1: Consider the functions

$$F_i(t) = \text{Im}\left[\overline{(z_i(t) - z_0)}(z_{i+1}(t) - z_0)\right], \quad i = 1, \ldots, n-1,$$
$$F_n(t) = \text{Im}\left[\overline{(z_n(t) - z_0)}(z_1(t) - z_0)\right],$$

where z_0 is the centroid (see Fig. 7.9). Then by the definition of a counterclockwise star formation and Lemma 7.2, it follows that $r_i(0) > 0$ and $0 < \alpha_i(0) < \pi$, $i = 1, 2, \ldots, n$. That means $F_i(0) > 0$ for all i by Lemma 7.1. We need to show that $F_i(t) > 0$ for all t, implying $r_i(t) > 0$ and $0 < \alpha_i(t) < \pi$ for all t.

For a proof by contradiction, suppose that some F_i (without loss of generality, say F_1) becomes zero and it does so for the first time at $t = t_1$. Not all F_i's are zero at t_1, for then the points would be collinear, by Lemma 7.1, which would be a contradiction, by Lemma 7.3. Suppose in fact that $F_2(t_1) > 0$. We therefore have

$$F_i(t) \geq 0, \quad \forall t \in [0, t_1], \; i = 1, \ldots, n$$
$$F_1(t_1) = 0,$$
$$F_2(t_1) > 0.$$

Taking the derivative along the trajectory of system (7.2) and noting that $\dot{z}_0 = 0$ (the centroid is stationary by Theorem 5.1), we have

$$
\begin{aligned}
\dot{F}_1 &= \frac{d}{dt} \operatorname{Im} \left[\overline{(z_1 - z_0)}(z_2 - z_0) \right] \\
&= \operatorname{Im} \left[\overline{\dot{z}_1}(z_2 - z_0) \right] + \operatorname{Im} \left[\overline{(z_1 - z_0)}\dot{z}_2 \right] \\
&= \operatorname{Im} \left[\overline{(z_2 - z_1)}(z_2 - z_0) \right] + \operatorname{Im} \left[\overline{(z_1 - z_0)}(z_3 - z_2) \right] \\
&= \operatorname{Im} \left[\overline{((z_2 - z_0) - (z_1 - z_0))}(z_2 - z_0) \right] + \operatorname{Im} \left[\overline{(z_1 - z_0)}((z_3 - z_0) - (z_2 - z_0)) \right] \\
&= 0 - F_1 + G_1 - F_1, \\
&= -2F_1 + G_1,
\end{aligned}
$$

where

$$G_1 = \operatorname{Im} \left[\overline{((z_1 - z_0)}(z_3 - z_0)) \right].$$

By the formula $F_1 = r_1 r_2 \sin(\alpha_1)$, the condition $F_1(t_1) = 0$ implies that at $t = t_1$ at least one of the following holds:

1. $\alpha_1 = \pi$ and $r_1, r_2 > 0$,
2. $\alpha_1 = 0$ and $r_1, r_2 > 0$,
3. $r_2 = 0$,
4. $r_1 = 0$.

Let us show each of these is impossible.

1. Suppose $\alpha_1(t_1) = \pi$ and $r_1(t_1), r_2(t_1) > 0$. On the one hand, $\alpha_1(t_1) = \pi$ implies that at $t = t_1$, the centroid z_0 lies on the line formed by z_1 and z_2 (see Fig. 7.14). On the other hand, since $F_2(t_1) > 0$ and $F_i(t_1) \geq 0$ for all i, it follows from Lemma 7.1 that at $t = t_1$, $\alpha_2 > 0$ and $\alpha_i \geq 0$ for all i (Fig. 7.14), which implies the centroid z_0 does not lie on the line formed by z_1 and z_2, a contradiction.

2. Suppose $\alpha_1(t_1) = 0$ and $r_1(t_1), r_2(t_1) > 0$. Observe that

$$G_1 = r_1 r_3 \sin (\alpha_1 + \alpha_2),$$

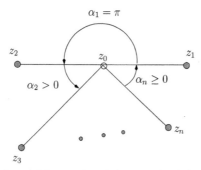

Fig. 7.14. Impossible case 1: $\alpha_1(t_1) = \pi$ and $r_1(t_1), r_2(t_1) > 0$.

so we have

$$G_1(t_1) = r_1(t_1)r_3(t_1)\sin(\alpha_1(t_1) + \alpha_2(t_1))$$
$$= r_1(t_1)r_3(t_1)\sin(\alpha_2(t_1))$$
$$= \frac{r_1(t_1)}{r_2(t_1)}F_2(t_1).$$

Hence $G_1(t_1) > 0$. By continuity, there exists t_0 ($t_0 < t_1$) such that $G_1(t) > 0$ for all $t \in [t_0, t_1]$. Also, by assumption, $F_1(t) > 0$ for $t \in [0, t_1)$. Hence,

$$\dot{F}_1 = -2F_1 + G_1 > -2F_1, \quad \text{for } t \in [t_0, t_1)$$

and therefore

$$F_1(t) > \exp\left(-2(t - t_0)\right)F_1(t_0) > 0, \quad \text{for } t \in [t_0, t_1).$$

Then by continuity of F_1, $F_1(t_1) > 0$, a contradiction.

3. $r_2(t_1) = 0$ is impossible since $F_2(t_1) > 0$.

4. Finally, suppose $r_1(t_1) = 0$, that is, $z_1(t_1)$ and z_0 coincide. As the pursuit curves are coordinate-free, without loss of generality, it is assumed that the coordinate system has the origin at z_0 and its negative real axis passes through the location $z_2(t_1)$ (see Fig. 7.15). Denote $r := r_2(t_1) > 0$. Then, in the coordinate system above, we have

$$\dot{z}_1(t_1) = z_2(t_1) - z_1(t_1) = z_2(t_1) = -r.$$

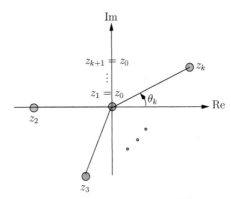

Fig. 7.15. Impossible case 4: $r_1(t_1) = 0$.

Since it is impossible to have all points except z_2 on the centroid z_0 at time $t = t_1$, there must be a $k > 2$ such that $r_k(t_1) \neq 0$ and $r_i(t_1) = 0$ for $i = k+1, \ldots, n$. (In other words, $z_1, z_n, z_{n-1}, \ldots, z_{k+1}$ are co-located at z_0 at $t = t_1$ while z_k is not.) Then, we claim that $\mathrm{Im}\,[z_k(t_1)] > 0$ as in Fig. 7.15. (To see this, suppose by contradiction that $\mathrm{Im}\,[z_k(t_1)] \leq 0$. Then by the fact that $F_i(t_1) \geq 0$ for all i, it follows that for any $i \in \{3, 4, \ldots, k-1\}$, $\mathrm{Im}\,[z_i(t_1)] \leq 0$. Moreover, notice that $\frac{1}{n}\mathrm{Im}\,[z_1(t_1) + z_2(t_1) + \cdots + z_n(t_1)] = \mathrm{Im}\,[z_0] = 0$, so we know for any $i \in \{3, 4, \ldots, k-1\}$, $\mathrm{Im}\,[z_i(t_1)] = 0$, which implies the n points are collinear at $t = t_1$. Then by Lemma 7.3, they are collinear for all time, a contradiction.)

Let $\theta_k(t_1) = \delta > 0$ (Fig. 7.15). By continuity of θ_k, there exists t_0 ($t_0 < t_1$) such that for $t \in [t_0, t_1]$, $\theta_k(t) > \delta/2$. Since the points are in a counterclockwise star formation until t_1 we have that $\theta_1(t) > \theta_k(t) > \delta/2$ for $t \in [t_0, t_1)$. Note that

$$z_1(t_1) - z_1(t_1 - h) = h\dot{z}_1(t_1) - O(h),$$

where $O(h)/h \to 0$ as $h \to 0$. Then we obtain

$$z_1(t_1 - h) = z_1(t_1) - h\dot{z}_1(t_1) + O(h) = rh + O(h).$$

In other words,

$$\mathrm{Re}\left[z_1(t_1 - h)\right] = rh + O(h)$$
$$\mathrm{Im}\left[z_1(t_1 - h)\right] = O(h).$$

Thus, for $h > 0$ sufficiently small, $\mathrm{Re}\left[z_1(t_1 - h)\right] > 0$ and $\theta_1(t_1 - h) > \delta$ (see Fig. 7.16). Furthermore, it is obtained that

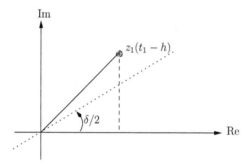

Fig. 7.16. Illustration for $\mathrm{Im}\left[z_1(t_1 - h)\right] > \mathrm{Re}\left[z_1(t_1 - h)\right]\tan \delta/2$.

$$\mathrm{Im}\left[z_1(t_1 - h)\right] > \mathrm{Re}\left[z_1(t_1 - h)\right]\tan \delta/2.$$

Combining these facts we then have

$$O(h) > (rh + O(h))\tan \delta/2.$$

Dividing by h and taking the limit as $h \to 0$ we obtain

$$\lim_{h \to 0} O(h)/h > r \tan \delta/2 > 0,$$

a contradiction.   ∎

7.4 Formation Control via State-Independent Graph

Instead of convergence to a point, can we get three robots to form a triangle (Fig. 7.17) by a distributed control strategy?

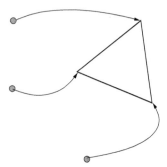

Fig. 7.17. Can we get three robots to form a triangle?

Continue with the cyclic pursuit strategy, but now we allow each robot to pursue a *virtual displacement* of the next robot:

$$\dot{z}_1 = (z_2 + d_1) - z_1,$$
$$\dot{z}_2 = (z_3 + d_2) - z_2,$$
$$\dot{z}_3 = (z_1 + d_3) - z_3.$$

The vector form is

$$\dot{z} = Az + d \tag{7.3}$$

where

$$z = \begin{bmatrix} z_1 \\ z_2 \\ z_3 \end{bmatrix}, \quad A = \begin{bmatrix} -1 & 1 & 0 \\ 0 & -1 & 1 \\ 1 & 0 & -1 \end{bmatrix}, \quad d = \begin{bmatrix} d_1 \\ d_2 \\ d_3 \end{bmatrix}.$$

Let **1** again denote the vector of all 1s: an eigenvector of A for the zero eigenvalue. Pre-multiplying (7.3) by $\mathbf{1}^T$ leads to

$$\frac{d(\mathbf{1}^T z)}{dt} = \mathbf{1}^T d.$$

Thus if $\mathbf{1}^T d \neq 0$, that is, if the centroid of the points d_1, d_2, d_3 is not at the origin, then the centroid of the robots moves off to infinity. To avoid this, we assume that $\mathbf{1}^T d = 0$. Then d lies in the stable eigenspace of A. Let c denote the unique vector satisfying $Ac + d = 0$. Then (7.3) can be written as

$$\frac{d}{dt}(z(t) - c) = A(z(t) - c).$$

From Theorem 5.1, the centroid of the points $z_1(t) - c_1, z_2(t) - c_2, z_3(t) - c_3$ (c_1, c_2, c_3 are the components of c) is stationary and every $z_i(t) - c_i$ converges to this centroid. It is equivalent to say that every robot i converges to this centroid displaced by c_i and thus a triangle formation is achieved.

Now let us study this for a general formation and n robots in the place.

First we introduce the description of a geometric formation. A *formation* is defined by a set of n complex numbers, i.e, $c = (c_1, \ldots, c_n)$. It should be cleared up that this description is independent of the choice of complex plane since only the relative information will be used. In other words, if there exist a rotation ψ and a translation $\xi \in \mathbb{C}$ such that $c_i' = e^{j\psi} c_i + \xi$ for all i, then c' and c represent the same formation. An illustrative example is given in Fig. 7.18. Both $c = (c_1, c_2, c_3)$ and $c' = (c_1', c_2', c_3')$ in Fig. 7.18 describe the same triangle formation.

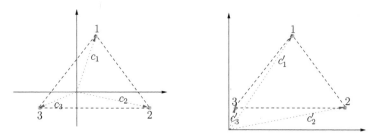

Fig. 7.18. Description of triangle formation.

Next, we present formal definitions to explain what we mean by saying that the robots achieve a formation.

Definition 7.2. *A group of n robots achieve a formation $c = (c_1, \ldots, c_n)$ if for any initial state $z_i(0), i = 1, \ldots, n$, there exist a rotation ψ and a translation $\xi \in \mathbb{C}$ such that for all i, $z_i(t)$ converges to $e^{j\psi} c_i + \xi$.*

Suppose each robot i can obtain the relative positions of its neighbor robots that are defined in the information flow graph $\mathcal{G}(t)$, namely,

$$z_j - z_i, \text{ if } j \in \mathcal{N}_i(t),$$

where $\mathcal{N}_i(t)$ is the set of neighbors at t. In this section, we consider the case that the information flow graph $\mathcal{G}(t)$ for the n robots are state-independent. Then the objective is to devise the controller for each robot using its available information so that the group of robots can form a desired geometric formation.

Here comes our main result.

Theorem 7.2. *Assume the information flow graph $\mathcal{G}(t)$ is uniformly quasi-strongly connected. Then the n robots (7.1) achieve a formation $c = (c_1, \ldots, c_n)$ with the distributed control*

$$u_i = \sum_{j \in \mathcal{N}_i(t)} (z_j - z_i) - \sum_{j \in \mathcal{N}_i(t)} (c_j - c_i). \tag{7.4}$$

Proof: Considering the distributed control law (7.4), we obtain the closed-loop system in a matrix form

$$\dot{z}(t) = A_{\mathcal{G}(t)}(z(t) - c),$$

where z is the aggregate state of n robots and $A_{\mathcal{G}(t)}$ is negative Laplacian matrix corresponding to $\mathcal{G}(t)$.

Introduce new coordinate variables $\bar{z}_i = z_i - c_i$ $(i = 1, \ldots, n)$. Then we have

$$\dot{\bar{z}}(t) = A_{\mathcal{G}(t)} \bar{z}(t).$$

Thus, from Theorem 5.6, it follows that the system achieves state agreement. In other words, for any initial state there is a ξ such that every $\bar{z}_i(t)$ converges to ξ, which implies, for all i, $z_i(t)$ converges $c_i + \xi$. Hence, the conclusions follows by Definition 7.2. ∎

Looking at the control law (7.4), it can be noticed that the first term is the sensed information available to robot i and the second term is the formation vectors also available to robot i, so it is a decentralized controller. By the way, in order to implement the distributed control law (7.4) in the local frame of each robot and to achieve the desired formation, it is required that all local frames of n robots should have the same orientation.

Here we present two simulations. In Fig. 7.19, we simulate this control strategy on six robots to achieve an equilateral triangle formation described by

$$c = (0, 5 + j5\sqrt{3}, 10 + j10\sqrt{3}, 15 + j5\sqrt{3}, 20, 10).$$

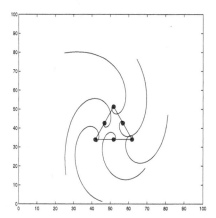

Fig. 7.19. Achieving equilateral triangle formation.

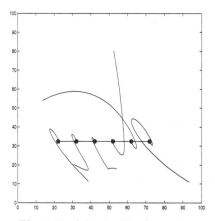

Fig. 7.20. Achieving line formation.

And in Fig. 7.20, the desired formation is a line formation described by

$$c = (0, 10, 20, 30, 40, 50).$$

Suppose there is an external reference velocity for all the n robots and consider the modified distributed control law

$$u_i = \sum_{j \in \mathcal{N}_i(t)} (z_j - z_i) - \sum_{j \in \mathcal{N}_i(t)} (c_j - c_i) + v_{ref}. \qquad (7.5)$$

With almost similar analysis, we obtain the following result.

Corollary 7.1. *Assume the information flow graph $\mathcal{G}(t)$ is uniformly quasi-strongly connected. Then the n robots with the distributed control (7.5) eventually move in formation.*

The external reference velocity v_{ref} here determines the speed and the orientation of the group of robots moving as a whole. The behaviors of moving in formation for a group of six robots are simulated in Fig. 7.21 and Fig. 7.22,

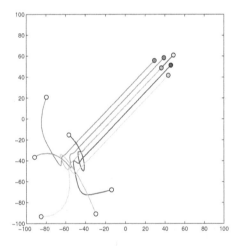

Fig. 7.21. Moving in formation (fixed information flow graph).

where the external reference velocity is given by $v_{ref} = 1 + j$ and the desired formation is given by

$$c = (0, 9.66 + j2.59, 19.32 + j5.18, 16.73 - j4.48, 14.14 - j14.14, 7.07 - j7.07).$$

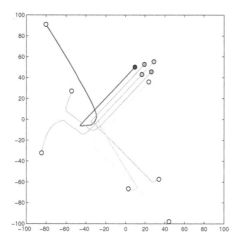

Fig. 7.22. Moving in formation (dynamic information flow graph).

The information flow graph corresponding to the simulation in Fig. 7.21 is fixed, while the information flow graph corresponding to the simulation in Fig. 7.22 is time-varying. It can also be noticed that the trajectories in Fig. 7.22 have several places sharp angled.

7.5 Rendezvous via State-Dependent Graph

In this section, we take a more realistic sensor model and assume each robot can see only a given distance r. The set of visible neighbors of robot i is

$$\mathcal{N}_i(z) := \{j : \|z_j - z_i\| \le r\}.$$

Then there is an edge in the information flow graph from i to j if and only if $j \in \mathcal{N}_i(z)$, and therefore the information flow graph is a function of state z, denoted by $\mathcal{G}(z)$. First, we observe that the information flow graph defined here is bidirectional because the visibility radii are assumed to be same for all robots. Second, the information flow graph is state-dependent, not time-dependent. If a control law is given and if the state $z(t)$ evolves uniquely from $z(0)$, the information flow graph is then a function of time, $\mathcal{G}(z(t))$.

Assume again, each robot can only measure the relative position of its neighbors. Consider for example six robots, with omnidirectional cameras of identical ranges, positioned at $t = 0$ as in Fig. 7.23. The discs show the fields of view for robots 1 and 2. Since each camera has the same range, the information flow graph is bidirectional—Fig. 7.24. Thus the neighbor sets at $t = 0$ are

$$\mathcal{N}_1 = \{2\}, \; \mathcal{N}_2 = \{1,4\}, \; \mathcal{N}_3 = \{4\}, \; \mathcal{N}_4 = \{2,3,5\}, \; \mathcal{N}_5 = \{4,6\}, \; \mathcal{N}_6 = \{5\},$$

and the measurement for robot 2, for instance, is $(z_1 - z_2), \; (z_4 - z_2)$.

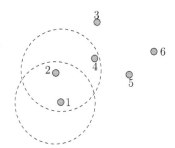

Fig. 7.23. Six robots; the fields of view of robots 1 and 2 are shown.

Fig. 7.24. Information flow graph.

We deal with the rendezvous problem in this section. Here is our definition for robots rendezvous[1].

Definition 7.3. *A group of n robots* rendezvous *if for any initial state $z_i(0), i = 1, \ldots, n$, there exists a point ζ such that for all i, $z_i(t)$ converges to ζ.*

Let us first consider the linear distributed control

$$u_i = \sum_{j \in \mathcal{N}_i(z)} (z_j - z_i)$$

and then we have the following closed-loop overall system of n robots

$$\dot{z} = A_{\mathcal{G}(z)} z,$$

where $A_{\mathcal{G}(z)}$ is the negative Laplacian matrix corresponding to the graph $\mathcal{G}(z)$. Theorem 5.6 suggests that the system achieves state agreement (robots rendezvous) if and only if the information flow graph $\mathcal{G}(z(t))$ is uniformly quasi-strongly connected. However, as the information flow graph evolves depending on the trajectory $z(t)$, the condition becomes non-checkable. The information flow graph may be uniformly quasi-strongly connected for some initial conditions and may not for some other initial conditions. To understand the point, we run a simulation of twenty robots using the above linear distributed control law three times with the same initial locations, but different sensing ranges. Fig. 7.25 shows the initial locations generated randomly. Fig. 7.26, 7.27, and 7.28 show the trajectories of the twenty robots for the sensing range $r = 30$, 25, and 50, respectively. From the simulation, we observe that for the case of $r = 25$ the twenty robots converge to two locations rather than rendezvous, and for the other two cases ($r = 30$, 50), they do rendezvous.

Clearly, if some robots are initialized so far away from the rest that they never come to their fields of view, then rendezvous never occurs. Mathematically, this corresponds to the situation where $\mathcal{G}(z(0))$ is not quasi-strongly connected. On second thought, one may ask what if the information flow graph is quasi-strongly connected initially. Unfortunately, the linear distributed control law seems not good enough to solve the rendezvous problem even though it is quasi-strongly connected initially because some already established link may be dropped as the

---

[1] The Oxford English Dictionary definition of *to rendezvous* is "to assemble at a place previously appointed; also generally, to assemble, come together, meet." Without a supervisor or global coordinates known to all, the robots are incapable of assembling at a place previously appointed. The best they can do is to meet at some not pre-specified place.

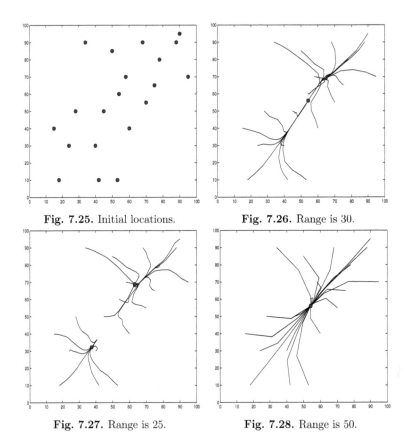

Fig. 7.25. Initial locations.

Fig. 7.26. Range is 30.

Fig. 7.27. Range is 25.

Fig. 7.28. Range is 50.

system evolves. An example is given in Fig. 7.29, where it can be observed that the link between robot 2 and 3 will be dropped immediately when using the linear distributed control above.

Next we will try to find a nonlinear distributed control

$$u_i = f^i \left((z_j - z_i)|_{j \in \mathcal{N}_i(z)} \right) \tag{7.6}$$

to solve the rendezvous problem. But first we assume that $\mathcal{G}(z(0))$ is quasi-strongly connected. And then, we wish the nonlinear distributed control guarantees $\mathcal{G}(z(t))$ not losing this property in the future even though the control

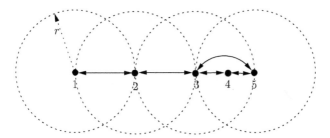

Fig. 7.29. Extreme situation.

may cause link changes in $\mathcal{G}(z(t))$. Intuitively, the control u_i should make the maximum distance between robot i and its neighbor robots non-increasing. This observation leads to the following result.

Let $\mathcal{I}_i(z)$ denote the set of neighbor robots $j \in \mathcal{N}_i(z)$ that have maximum distance from robot i, i.e.,

$$\mathcal{I}_i(z) = \{j \in \mathcal{N}_i(z) : \|z_j - z_i\| \geq \|z_k - z_i\| \text{ for any } k \in \mathcal{N}_i(z)\}.$$

Lemma 7.4. *If, for all i, u_i satisfies*

$$\max_{j \in \mathcal{I}_i(z)} (z_i - z_j)^T u_i \leq 0, \tag{7.7}$$

then over time no links are dropped in the information flow graph.

Proof: For this proof it is more convenient to view a robot position z_i as a vector in $\mathbb{R}^2$ instead of a complex number. Let $V_{ij}(z)$ denote the distance-squared between two neighbor robots i and j, and let $V(z)$ denote the maximum distance-squared between any two neighbors:

$$V(z) = \max_i \max_{j \in \mathcal{N}_i(z)} \|z_i - z_j\|^2.$$

Let $\mathcal{I}(z)$ denote the set of pairs of indices where the maximum is attained; that is, $(i, j) \in \mathcal{I}(z)$ if and only if robots i and j are neighbors of maximum distance apart among all neighbors. Thus

$$V(z) = \max_{(i,j) \in \mathcal{I}(z)} V_{ij}(z).$$

We would like to show that $\frac{d}{dt}V(z(t)) \le 0$. Unfortunately, $V(z(t))$ is not differentiable. We will use some non-smooth analysis—the upper Dini derivative $D^+V(z(t))$ instead. Then, by Lemma B.2, it follows that

$$D^+V(z(t)) = \max_{(i,j)\in\mathcal{I}(z(t))} \frac{d}{dt}V_{ij}(z(t)).$$

In this way we get

$$
\begin{aligned}
D^+V(z(t)) &= \max_{(i,j)\in\mathcal{I}(z)} 2(z_i - z_j)^T(\dot{z}_i - \dot{z}_j) \\
&= \max_{(i,j)\in\mathcal{I}(z)} \left\{ 2(z_i - z_j)^T u_i + 2(z_j - z_i)^T u_j \right\} \\
&\le \max_{(i,j)\in\mathcal{I}(z)} 2(z_i - z_j)^T u_i + \max_{(i,j)\in\mathcal{I}(z)} 2(z_j - z_i)^T u_j.
\end{aligned}
$$

Then it follows from the condition (7.7) that

$$\max_{(i,j)\in\mathcal{I}(z)} (z_i - z_j)^T u_i \le 0 \quad \text{and} \quad \max_{(i,j)\in\mathcal{I}(z)} (z_j - z_i)^T u_j \le 0.$$

Thus,

$$D^+V(x(t)) \le 0, \quad \forall t \ge 0,$$

which means no links will be disconnected. ∎

From the lemma above, we know that if there exists a nonlinear distributed control (7.6) satisfying (7.7) then it preserves connectivity of the information flow graph. We now impose two additional conditions for the nonlinear distributed control (7.6) so that the class of nonlinear control laws can solve the rendezvous problem.

A1': u_i is continuous;

A2': $u_i \in \mathcal{T}(z_i, \mathcal{C}^i(z))$, where $\mathcal{C}^i(z) = \text{co}\{z_i, z_j|_{j\in\mathcal{N}_i(z)}\}$. In addition, $u_i \ne 0$ if z_i is $\mathcal{C}^i(z)$'s vertex and $\mathcal{C}^i(z)$ is not a singleton.

It should be noticed that when considering the nonlinear distributed control (7.6), the above two conditions are exactly the same as assumptions A1' and A2' in Section 6.3. Thus, we can call for the results in Section 6.3 as a tool.

The nonlinear distributed control (7.6) gives rise to the following closed-loop overall system

$$\dot{z} = f_{\mathcal{G}(z)}(z), \tag{7.8}$$

where $f_{\mathcal{G}(z)}$ is the vector field with components

$$f^1\left((z_j - z_1)|_{j \in \mathcal{N}_1(z)}\right),\ f^2\left((z_j - z_2)|_{j \in \mathcal{N}_2(z)}\right),\ \ldots,\ f^n\left((z_j - z_n)|_{j \in \mathcal{N}_n(z)}\right).$$

We will show in the following lemma that if the distributed control law u_i satisfies the condition (7.7) as well as conditions A1′ and A2′, a solution $z(t)$ to (7.8) exists and the robots rendezvous.

Theorem 7.3. *Suppose $\mathcal{G}(z(0))$ is quasi-strongly connected. If u_i satisfies (7.7) as well as A1′ and A2′, then the robots rendezvous.*

Proof: If u_i satisfies (7.7), then over time no links are dropped by Lemma 7.4. So if $\mathcal{G}(z(0))$ is fully connected, then $\mathcal{G}(z(t))$ is always fully connected, which means it is fixed for all time. Then the conclusion follows from Theorem 6.2.

If instead $\mathcal{G}(z(0))$ is not fully connected, then we claim that $\mathcal{G}(z(t))$ is dynamic but only switches for a finite number of times. To see this, suppose by contradiction that for all $t \geq 0$, $\mathcal{G}(z(t)) = \mathcal{G}(z(0))$. Then by Theorem 6.2, all the robots converge to a common location, which leads to a fully connected information flow graph at some time t, a contradiction. Hence, there is a $t_1 \geq 0$ such that $\mathcal{G}(z(t_1))$ has more arcs than $\mathcal{G}(z(0))$ because no arc will be dropped by Lemma 7.4. Repeating this argument a finite number of times eventually leads to the existence of t_i such that $\mathcal{G}(z(t_i))$ is fully connected, and thus, it is fixed after t_i. Then the conclusion follows from Theorem 6.2 by treating $\left(t_i, z(t_i)\right)$ as the initial condition. ∎

Now we turn to the *circumcentre control law* that actually satisfy conditions (7.7) as well as A1′ and A2′. The circumcentre control law was first proposed in [4]. Consider the example of six robots in Fig. 7.23 again. For that example, the circumcentre control law is defined as follows: Robot 1 has one neighbor, robot 2. Let c_1 denote the circumcentre of $\mathcal{C}^1 = \mathrm{co}\{z_1, z_2\}$—the center of the smallest circle containing $\mathcal{C}^1$. Then set $u_1 = c_1 - z_1$. (In Fig. 7.30, the little arrow is u_1 translated from the origin to z_1.) So robot 1 moves towards the center at $t = 0$: $\dot{z}_1 = c_1 - z_1$. Actually, in this case where robot 1 sees only robot 2, clearly $c_1 = (z_1 + z_2)/2$, so at $t = 0$

$$\dot{z}_1 = \frac{1}{2}(z_2 - z_1).$$

Similarly, let c_2 denote the circumcentre of the set $\mathcal{C}^2 = \{z_2, z_1, z_4\}$ and define $u_2 = c_2 - z_2$ (see Fig. 7.31). And so on.

These control laws can actually be implemented using onboard cameras, that is, relative positions, by translation. For example, for robot 2, the relative po-

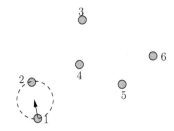

Fig. 7.30. Control law for robot 1.

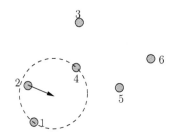

Fig. 7.31. Control law for robot 2.

sitions $\{z_1 - z_2, z_4 - z_2\}$ are sensed. Let $\tilde{\mathcal{C}}^2$ denote the convex hull of points $\{0, z_1 - z_2, z_4 - z_2\}$ (the translation of $\mathcal{C}^2$ by $-z_2$), and let c'_2 denote the circumcentre of $\tilde{\mathcal{C}}^2$. Then $u_2 = c_2 - z_2$ is equivalent $u_2 = c'_2$. Recall that the set $\tilde{\mathcal{C}}^2$ is actually depending on the state, namely, $\{z_1 - z_2, z_4 - z_2\}$, and so the circumcentre c'_2 is a function of z.

In general, we denote $\tilde{\mathcal{C}}^i(z)$ the translation of $\mathcal{C}^i(z)$ by $-z_i$ and write the circumcentre control law

$$u_i(z) = c'_i \left((z_j - z_i)|_{j \in \mathcal{N}_i(z)} \right),$$

which is of the form of the nonlinear distributed control (7.6). In this way, the robot's motions are governed by the coupled equations

$$\dot{z}_1 = u_1(z) = c_1' \left((z_j - z_1)|_{j \in \mathcal{N}_1(z)} \right)$$

$$\vdots$$

$$\dot{z}_n = u_n(z) = c_1' \left((z_j - z_n)|_{j \in \mathcal{N}_n(z)} \right).$$

Now let's present some properties of the circumcentre control law.

Lemma 7.5. *The circumcentre control law is continuous.*

Proof: Consider a set of points $\{p_1, p_2, \ldots, p_m\}$. Let $c(p_1, \ldots, p_m)$ be the circumcentre of the set $\mathcal{C} = \text{co}\{p_1, \ldots, p_m\}$. It is actually the unique point that minimizes the function

$$g(w) := \max_{p \in \{p_1, \ldots, p_m\}} \|w - p\|.$$

Interpreted geometrically, $c(p_1, \ldots, p_m)$ is the center of the smallest circle con-

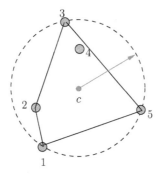

Fig. 7.32. Circumcentre of a set of points.

taining the set of points $\{p_1, \ldots, p_m\}$ (see Fig. 7.32). Furthermore, it can be easily shown that it lies in the polytope $\mathcal{C}$ but not at its vertices if the polytope is not a singleton. Thus,

$$c(p_1, \ldots, p_m) = \arg \min_{w \in \mathcal{C}} g(w).$$

Then, by Theorem E.1 (Berge's Maximum Theorem), the function $c(p_1, \ldots, p_m)$ is continuous with respect to its arguments. ∎

Lemma 7.6 ([20]). *The circumcentre control law is not Lipschitz continuous.*

Proof: Construct three points $\{p_1, p_2, p_3\}$ and their circumcentre c, and three perturbed points $\{p_1, p_2', p_3'\}$ and their circumcentre c' as in Fig. 7.33. Define the

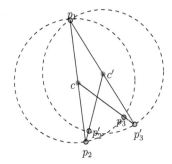

Fig. 7.33. Circumcentre and perturbed circumcentre.

vector

$$p = \begin{bmatrix} p_1 \\ p_2 \\ p_3 \end{bmatrix}, \quad p' = \begin{bmatrix} p_1 \\ p_2' \\ p_3' \end{bmatrix}.$$

We will show that the ration

$$\frac{\|c - c'\|}{\|p - p'\|}$$

is not bounded by a constant. This proves the mapping $p \mapsto c$ is not Lipschitz continuous.

Let the radii of the circles be 1 and define $x = \|c - c'\|$, $y = \|p_2 - p_2'\|$. Since p_1 did not move and $\|p_2 - p_2'\| = \|p_3 - p_3'\|$, we have

$$\|p - p'\| = \sqrt{2}y.$$

Now look at Define the angle θ as in Fig. 7.34. Then we have the length $\overline{p_1 q} = \cos\theta$, $\overline{qc} = 1 - \cos\theta$, so by Pythagoras on the small triangle qcc'

$$x^2 = (1 - \cos\theta)^2 + \sin^2\theta = 2(1 - \cos\theta),$$

and therefore $\overline{qc} = x^2/2$.

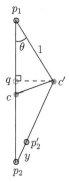

Fig. 7.34. Illustration for the proof of Lemma 7.6.

By Pythagoras again on the triangle qcc', the length of $\overline{qc'}$ equals $x\sqrt{1 - \frac{x^2}{4}}$. Finally, apply Pythagoras to triangle qp_2c':

$$(y+1)^2 = \left(1 + \frac{x}{2}\right)^2 + x^2\left(1 - \frac{x^2}{4}\right)^2 = 2x^2 + 1.$$

Thus we have

$$x = \sqrt{\frac{1}{2}y^2 + y},$$

and so

$$\frac{\|c - c'\|}{\|p - p'\|} = \frac{x}{\sqrt{2}y} = \frac{1}{2}\sqrt{1 + \frac{2}{y}},$$

which is not bounded. ∎

Theorem 7.4. *Suppose $\mathcal{G}(z(0))$ is quasi-strongly connected. Under the circum-centre control law, the robots rendezvous.*

Proof: First, by Lemma 7.5, u_i is continuous and so satisfies A1'.

Next, from the proof of Lemma 7.5, we know that $u_i = c'_i\left((z_j - z_i)|_{j \in \mathcal{N}_i(z)}\right)$ lies in $\tilde{\mathcal{C}}^i(z)$ but not at its vertices if $\tilde{\mathcal{C}}^i(z)$ is not a singleton. That means, $u_i \in \mathcal{T}\left(0, \tilde{\mathcal{C}}^i(z)\right)$, and if 0 is a vertex of $\tilde{\mathcal{C}}^i(z)$ that is not a singleton then $u_i \neq 0$. Note that $\tilde{\mathcal{C}}^i(z)$ is just a translation of $\mathcal{C}^i(z)$. So $\mathcal{T}\left(0, \tilde{\mathcal{C}}^i(z)\right)$ is exactly the same as $\mathcal{T}\left(0, \mathcal{C}^i(z)\right)$. Moreover, 0 is a vertex of $\tilde{\mathcal{C}}^i(z)$ if and only if z_i is a vertex of $\mathcal{C}^i(z)$. Thus u_i satisfies A2'.

Finally, we show that u_i satisfies (7.7). This can be seen from geometry. If $u_i = 0$, then it is trivia. If $u_i \neq 0$, then the picture is as in Fig. 7.35. The

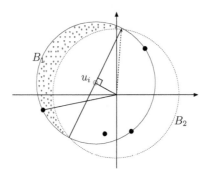

Fig. 7.35. Illustration for the proof of Theorem 7.4.

solid circle B_1 is the smallest circle enclosing the points $\{0, (z_j - z_i)|_{j \in \mathcal{N}_i(z)}\}$. Construct the line as shown through the point u_i perpendicular the vector u_i and construct the dotted circle B_2 that is centered at the origin and goes through the intersection points of B_1 and the line just constructed. We know that if there are some $(z_j - z_i)$ in the closed shaded area, then one of them achieves the maximal distance from the origin among all points. On the other hand, there is at least one $j \in \mathcal{N}_i(z)$ such that $(z_j - z_i)$ is in the closed semicircle of B_1, since otherwise it is not the smallest circle enclosing all the points. Hence, if $j \in \mathcal{I}_i(z)$ then $(z_j - z_i)$ must lie in the closed shaded area, which means the angle between u_i and $(z_j - z_i)$ is less than $\pi/2$. Hence,

$$\max_{j \in \mathcal{I}_i(z)} (z_i - z_j)^T u_i \leq 0$$

and (7.7) is satisfied.   ∎

Though the analysis in this section is for the rendezvous problem in a plane, it is also true for any dimension.

7.6 Notes and Discussion

We take a look at the rendezvous problem and its extensions in a few other contexts.

Discrete-Event Robots

Versions of the robot rendezvous problem have been studied extensively in computer science (where it is usually called the *gathering problem*) [28, 43, 48]. Let us look at [43] as an interesting example.

Each robot is viewed as a point in the plane. The robots have limited visibility: Each can see only the other robots within a fixed radius. Moreover, the robots are modeled as asynchronous discrete-event systems (Fig. 7.36) having four possible states: *Wait*, that is, not moving and idle; *Look*, during which the

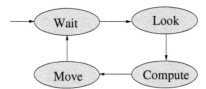

Fig. 7.36. Discrete-event robot.

robot senses the relative positions of the other robots within its field of view; *Compute*, during which it computes its next move; and *Move*, during which it moves at some pre-determined speed to its computed destination. There are soft timing assumptions, such as, a robot can be in Wait for only a finite period of time.

The robots have local coordinate frames and these are assumed to have a common orientation, e.g., they may each have a compass; see Fig. 7.37. The paper proposes the following control law, in the form of four if-then rules:

1. If in the Look state a robot sees another robot to its left or vertically above, then it does not move.

2. If a robot sees robots only below on its vertical axis, then it moves down toward the nearest robot.

3. If a robot sees robots only to its right, then it moves horizontally toward the vertical axis of the nearest robot.

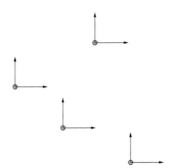

Fig. 7.37. Common orientation.

4. If a robot sees robots both below on its vertical axis and on its right, then it computes a certain destination point and performs a diagonal move down and to the right.

It is proved that, assuming the initial visibility graph is connected, the robots rendezvous after a finite number of events. For example, starting as in Fig. 7.37, the lower-right robot will not move, and the other three will become collocated with it. The proof is quite complicated, because, although each robot goes through a sequence of event cycles Wait-Look-Compute-Move, the robots are entirely unsynchronized, and so a robot may start to move before another has finished moving.

Discrete-Time Robots

Let us look in more detail at the model of Jadbabaie et al. [58]. For simplicity, suppose there are two boids, each a neighbor of the other. They move at unit speed with heading angles θ_1, θ_2 with respect to the global frame. The model in [58] is

$$\theta_1(k+1) = u_1(k) = \frac{1}{2}[\theta_1(k) + \theta_2(k)]$$
$$\theta_2(k+1) = u_2(k) = \frac{1}{2}[\theta_1(k) + \theta_2(k)].$$

The heading angles converge in just one time step. It may not be obvious, but these steering laws are feasible by onboard sensors. Rewrite the equations as

$$\theta_1(k+1) - \theta_1(k) = \frac{1}{2}[\theta_2(k) - \theta_1(k)]$$
$$\theta_2(k+1) - \theta_2(k) = \frac{1}{2}[\theta_1(k) - \theta_2(k)].$$

Thus the heading angles can be updated from the measured relative heading angles.

Cortés et al. [30] and Lin et al. [75] take the discrete-time robot model

$$z_i(k+1) = z_i(k) + u_i(k),$$

that is, the position update $z_i(k+1) - z_i(k)$ is directly controllable. If only local onboard cameras are available, then again u_i must be a function of the relative positions $(z_j - z_i)$, $j \in \mathcal{N}_i(z)$. The circumcentre control law is applied in [30] to solve the rendezvous problem. Lin et al. propose properties of a distributed control law and prove rendezvous under these properties with the assumption of graph connectivity.

From our point of view of onboard sensors and distributed control, there is an interesting point about discrete-time models. Consider, again for simplicity, just two robots, each the neighbor of the other. Suppose the robots head for each other according to the equations

$$\dot{z}_1 = u_1 = z_2 - z_1$$
$$\dot{z}_2 = u_2 = z_1 - z_2.$$

Now suppose the onboard controllers are digital: The sensed signals (z_2-z_1), (z_1-z_2) are sampled via a periodic sampler S with sampling period T, then converted from discrete time to continuous time via a zero-order holder H. See Fig. 7.38. Then the model at the sampling instants is

$$z_1[(k+1)T] = z_1(kT) + T[z_2(kT) - z_1(kT)]$$
$$z_2[(k+1)T] = z_2(kT) + T[z_1(kT) - z_2(kT)].$$

Note that it requires the two digital controllers in Fig. 7.38 to be synchronized somehow, by a centralized clock and communication system. So the system is not really distributed. Lack of synchroneity would lead to jitter, which could alternatively be modeled. (It is interesting to note that the above sampled-data system is unstable for large enough T.)

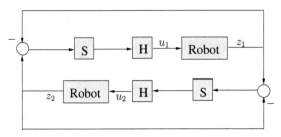

Fig. 7.38. Two robots with digital control.

8

Collective Kinematic Unicycles

Now that we have derived local control strategies for point-mass robots to achieve certain formation tasks, we may ask: Can we design a cooperative formation control system for a group of more complicated mobile robots—unicycle-type robots? As a fact again, the information flow topology among the group of unicycle-type robots is essential for the formation control problem. The goal of this chapter is to derive the conditions for the feasibility of formations.

In this chapter, we consider a group of robots which are identical and of kinematic unicycle model. Each one is equipped with an onboard sensor, by which it can measure relative displacements to certain neighbors. Information flow among the group of unicycles is again modeled by an information flow graph, where a link from node j to node i indicates that robot i can sense the position of robot j—but only with respect to the local coordinate frame of robot i. In addition, we assume in this chapter that the information flow graph is static—the dynamic case, where ad hoc links can be established or dropped, is harder and is still an open problem. Particularly, three subproblems, namely, rendezvous, row straightening, and pattern formation, are explored. Constructive local control is devised to achieve global asymptotic convergence of a group of unicycles to the desired formation in term of its shape.

8.1 Unicycle and Nonholonomic Constraint

Unicycle is probably the simplest model of a wheeled vehicle. As shown in Fig. 8.1, its kinematic model is given by

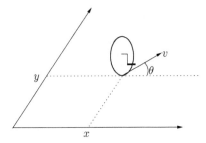

Fig. 8.1. Unicycle.

$$\begin{cases} \dot{x} = v\cos(\theta), \\ \dot{y} = v\sin(\theta), \\ \dot{\theta} = \omega, \end{cases} \tag{8.1}$$

where $z = (x,y) \in \mathbb{R}^2$ is the point the unicycle touches the ground, $\theta \in \mathbb{R}$ is the orientation, and the inputs v and ω are the forward and angular speeds, respectively.

The unicycle is kinematically equivalent to a mechanically very simple, and therefore common, *differential drive* wheeled vehicle such as shopping cart, which is also called *Hilare-type mobile robot* in the robotic society. One Hilare-type mobile robot is schematically depicted in Fig. 8.2: two independently actuated fixed

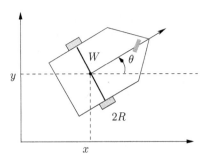

Fig. 8.2. Hilare-type mobile robot.

rear wheels and one small free moving front castor wheel to keep the balance. We denote the center position of the robot with $z = (x,y)$ and its orientation with θ. We also denote the rotation speed of its two wheel with v_r and v_l, respectively, and the radius of the wheel with R and the width between the wheels with W. Then its kinematic model is

$$\begin{cases} \dot{x} = \frac{R}{2}(v_r + v_l)\cos(\theta), \\ \dot{y} = \frac{R}{2}(v_r + v_l)\sin(\theta), \\ \dot{\theta} = \frac{R}{2W}(v_r - v_l). \end{cases}$$

The system above is equivalent to (8.1) through the following invertible coordinate transformation:

$$\begin{bmatrix} v \\ \omega \end{bmatrix} = \begin{bmatrix} \frac{R}{2} & \frac{R}{2} \\ \frac{R}{2W} & -\frac{R}{2W} \end{bmatrix} \begin{bmatrix} v_r \\ v_l \end{bmatrix}.$$

Thereafter, we call both unicycles without causing confusion.

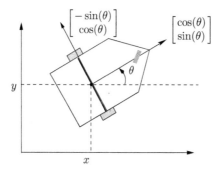

Fig. 8.3. Nonholonomic constraint.

The unicycle can move only in the direction it is heading (see Fig. 8.3). That is, there is no-slip condition:

$$\begin{bmatrix} \dot{x} \\ \dot{y} \end{bmatrix} \perp \begin{bmatrix} -\sin(\theta) \\ \cos(\theta) \end{bmatrix}$$

i.e.,

$$-\dot{x}\sin(\theta) + \dot{y}\cos(\theta) = 0.$$

This is called *nonholonomic constraint*.

More generally, consider a mechanical system with position vector q and velocity vector $\dot{q}$. a position constraint of the form

$$h(q) = 0$$

is called a *holonomic constraint* . An example is given in Fig. 8.4. Two point masses are connected by a massless rigid rod of length L. Let $q_1, q_2 \in \mathbb{R}^3$ be the

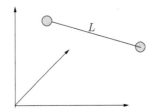

Fig. 8.4. Holonomic constraint.

positions of the two masses. Then the holonomic constraint is

$$\|q_1 - q_2\| - L = 0.$$

Thus a holonomic constraint restricts the motion to a surface in position space and reachability in configuration space is not possible.

A constraint of the form

$$h(q)\dot{q} = 0$$

that can not be integrated into a position constraint is called a *nonholonomic constraint*. Unicycle is an example:

$$\begin{bmatrix} -\sin(\theta) & \cos(\theta) & 0 \end{bmatrix} \begin{bmatrix} \dot{x} \\ \dot{y} \\ \dot{\theta} \end{bmatrix} = 0.$$

Let q denote the state (x, y, θ) of a unicycle. Then the unicycle model can be rewritten as

$$\dot{q} = v g_1(q) + \omega g_2(q),$$

where

$$g_1(q) = \begin{bmatrix} \cos(\theta) \\ \sin(\theta) \\ 0 \end{bmatrix}, \quad g_2(q) = \begin{bmatrix} 0 \\ 0 \\ 1 \end{bmatrix}.$$

This is of the form

$$\dot{q} = g_0(q) + u_1 g_1(q) + \cdots + u_m g_m(q),$$

which is called an *affine control system*, since for each q the mapping

$$u = (u_1, \ldots, u_m) \mapsto g_0(q) + u_1 g_1(q) + \cdots + u_m g_m(q) : \mathbb{R}^m \to \mathbb{R}^n$$

is an affine function (translate of a linear function). The control-less term $g_0(q)$ is called the *drift*. Thus the unicycle is drift-free: There is no motion when the inputs are zero. A unicycle on a moving flatbed truck would have a drift term.

8.2 Controllability and Stabilizability

In this section, we first introduce some geometric control theory ([62,97]) and see that the unicycle is a controllable nonlinear system. Next we present Brockett's necessary condition relating to the question of stabilizability of a nonlinear system and see that the unicycle cannot be stabilizable by a continuous, memoryless state feedback.

Let $\mathcal{F}(\mathbb{R}^n)$ denote the space of all smooth functions $\mathbb{R}^n \to \mathbb{R}^n$. E.g., unicycle:

$$q = \begin{bmatrix} x \\ y \\ \theta \end{bmatrix}, \ g_1(q) = \begin{bmatrix} \cos(\theta) \\ \sin(\theta) \\ 0 \end{bmatrix}, \ g_2(q) = \begin{bmatrix} 0 \\ 0 \\ 1 \end{bmatrix}, \ g_1, g_2 \in \mathcal{F}(\mathbb{R}^3).$$

The space $\mathcal{F}(\mathbb{R}^n)$ is a vector space over the field $\mathbb{R}$. It is an infinite-dimensional since it does not have a finite basis. For $f, g \in \mathcal{F}(\mathbb{R}^n)$, their *Lie bracket* is defined to be

$$[f, g](q) = \frac{\partial g}{\partial q} f(q) - \frac{\partial f}{\partial q} g(q).$$

Again, consider the unicycle as an example:

$$\frac{\partial g_1}{\partial q} = \begin{bmatrix} 0 & 0 & -\sin(\theta) \\ 0 & 0 & \cos(\theta) \\ 0 & 0 & 0 \end{bmatrix}, \quad \frac{\partial g_2}{\partial q} = \begin{bmatrix} 0 & 0 & 0 \\ 0 & 0 & 0 \\ 0 & 0 & 0 \end{bmatrix},$$

$$[g_1, g_2](q) = \frac{\partial g_2}{\partial q} g_1(q) - \frac{\partial g_1}{\partial q} g_2(q) = \begin{bmatrix} \sin(\theta) \\ -\cos(\theta) \\ 0 \end{bmatrix}.$$

Thus $[f, g] \in \mathcal{F}(\mathbb{R}^n)$ and so $\mathcal{F}(\mathbb{R}^n)$ is closed under Lie bracket. So $\mathcal{F}(\mathbb{R}^n)$ is a vector space and also it has a "product" operation defined on it; this makes it an *algebra*. Because the product operation is taking Lie bracket, we say $\mathcal{F}(\mathbb{R}^n)$ is a *Lie algebra*.

Returning to the unicycle, consider the two vectors $\{g_1, g_2\}$ in $\mathcal{F}(\mathbb{R}^3)$. Their span over $\mathbb{R}$, that is, the set of all linear combinations

$$c_1 g_1 + c_2 g_2, \quad c_1, c_2 \in \mathbb{R},$$

denoted $\mathrm{span}_{\mathbb{R}}\{g_1, g_2\}$, is a two-dimensional subspace of the infinite-dimensional vector space $\mathcal{F}(\mathbb{R}^3)$. However, $\mathrm{span}_{\mathbb{R}}\{g_1, g_2\}$ is not a subalgebra, because it is not closed under Lie bracket, i.e., $[g_1, g_2] \notin \mathrm{span}_{\mathbb{R}}\{g_1, g_2\}$:

$$\begin{bmatrix} \sin(\theta) \\ -\cos(\theta) \\ 0 \end{bmatrix} \neq c_1 \begin{bmatrix} \cos(\theta) \\ \sin(\theta) \\ 0 \end{bmatrix} + c_2 \begin{bmatrix} 0 \\ 0 \\ 1 \end{bmatrix}.$$

Let us add this extra function to get $\mathrm{span}_{\mathbb{R}}\{g_1, g_2, [g_1, g_2]\}$, a three-dimensional vector subspace of $\mathcal{F}(\mathbb{R}^3)$. It is closed under Lie bracket:

$$[g_1, [g_1, g_2]] = 0, \quad [g_2, [g_1, g_2]] = g_1.$$

It is therefore a Lie algebra itself. We denote $\mathrm{span}_{\mathbb{R}}\{g_1, g_2, [g_1, g_2]\}$ by $\mathrm{Lie}(\{g_1, g_2\})$, the *Lie algebra generated by* $\{g_1, g_2\}$.

More generally, let $\mathcal{S}$ be a subset of $\mathcal{F}(\mathbb{R}^n)$. For example, $\mathcal{S} = \{g_1, \ldots, g_m\}$, a finite set. The Lie algebra generated by $\mathcal{S}$, denoted by $\mathrm{Lie}(\mathcal{S})$, is the smallest subspace of $\mathcal{F}(\mathbb{R}^n)$ with two properties: It contains $\mathcal{S}$ and it is closed under Lie bracket, i.e.,

$$(\forall f, g \in \mathrm{Lie}(\mathcal{S}))\ [f, g] \in \mathrm{Lie}(\mathcal{S}).$$

For each fixed $q \in \mathbb{R}^n$ and $f \in \mathcal{F}(\mathbb{R}^n)$, $f(q)$ is a vector in $\mathbb{R}^n$. Again, for $\mathcal{S}$ a subset of $\mathcal{F}(\mathbb{R}^n)$, define

$$\mathrm{Lie}_q(\mathcal{S}) = \{f(q) : \ f \in \mathrm{Lie}(\mathcal{S})\}.$$

Thus $\mathrm{Lie}_q(\mathcal{S})$ is a vector subspace of $\mathbb{R}^n$.

For the unicycle,

$$g_1(q) = \begin{bmatrix} \cos(\theta) \\ \sin(\theta) \\ 0 \end{bmatrix}, \quad g_2(q) = \begin{bmatrix} 0 \\ 0 \\ 1 \end{bmatrix}, \quad [g_1, g_2](q) = \begin{bmatrix} \sin(\theta) \\ -\cos(\theta) \\ 0 \end{bmatrix}$$

$$\mathrm{Lie}(\{g_1, g_2\}) = \mathrm{span}_{\mathbb{R}}\{g_1, g_2, [g_1, g_2]\}$$

$$\mathrm{Lie}_q(\{g_1, g_2\}) = \{f(q) : \ f \in \mathrm{Lie}(\{g_1, g_2\})\}$$

$$= \mathrm{span}_{\mathbb{R}}\{g_1(q), g_2(q), [g_1, g_2](q)\}$$

$$= \mathbb{R}^3, \ \forall q.$$

This is an interesting linear example. Let $b \in \mathbb{R}^n$, $A \in \mathbb{R}^{n \times n}$, and define

$$f(x) = b, \quad g(x) = Ax.$$

Of course, these are smooth, so $f, g \in \mathcal{F}(\mathbb{R}^n)$; f is a constant function, g is a linear function. What is the Lie algebra, $\mathrm{Lie}(\{f, g\})$, generated by the two functions? Note that

$$[f, g](x) = \frac{\partial g}{\partial x} f(x) - \frac{\partial f}{\partial x} g(x) = Ab$$

$$[[f, g], g](x) = A^2 b$$

and so on. It follows that $\mathrm{Lie}(\{f, g\})$ equals the span over $\mathbb{R}$ of the functions

$$Ax, b, Ab, \ldots, A^{n-1}b.$$

(Terminates by Cayley-Hamilton.) Thus $\mathrm{Lie}(\{f, g\})$ is a finite-dimensional subspace of $\mathcal{F}(\mathbb{R}^n)$. For $x = 0$, $\mathrm{Lie}_x(\{f, g\})$ is the finite-dimensional subspace

$$\mathrm{span}_{\mathbb{R}}\{b, Ab, \ldots, A^{n-1}b\}$$

of $\mathbb{R}^n$, the controllable subspace of the pair (A, b).

Now let us get some ideas why the Lie algebra is important in nonlinear controllability.

Consider the 2-input system, like the unicycle:

$$\dot{q} = u_1 g_1(q) + u_2 g_2(q).$$

Starting at $q_0 = q(0)$, what states are reachable after a brief time interval? By taking

$$u_1(t) = 1, \quad u_2(t) = 0$$

we can head off in the direction $g_1(q_0)$. Likewise in the direction $g_2(q_0)$. What other directions? Let use consider in particular the following control actions (Fig. 8.5):

$$u_1(t) = 1, \quad u_2(t) = 0, \quad \forall t \in [0, \varepsilon)$$
$$u_1(t) = 0, \quad u_2(t) = 1, \quad \forall t \in [\varepsilon, 2\varepsilon)$$
$$u_1(t) = -1, \quad u_2(t) = 0, \quad \forall t \in [2\varepsilon, 3\varepsilon)$$
$$u_1(t) = 0, \quad u_2(t) = -1, \quad \forall t \in [3\varepsilon, 4\varepsilon).$$

Then where is the state at $t = 4\varepsilon$? Consider to use the following formula: For

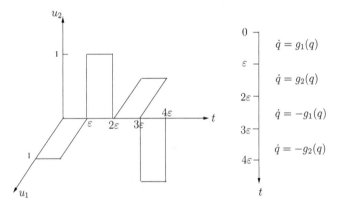

Fig. 8.5. Piecewise constant actions.

$\dot{q} = f(q)$, we have $\ddot{q} = \frac{\partial f}{\partial q}\dot{q} = \frac{\partial f}{\partial q}f$ and

$$q(t + \varepsilon) = q(t) + \varepsilon\dot{q}(t) + \tfrac{1}{2}\varepsilon^2\ddot{q}(t) + O(\varepsilon^3)$$
$$= q(t) + \varepsilon f(q(t)) + \tfrac{1}{2}\varepsilon^2\tfrac{\partial f}{\partial q}(q(t))f(q(t)) + O(\varepsilon^3).$$

In the following derivation, it is convenient to drop the evaluation argument q_0, like this:

Step 1. $\dot{q} = g_1(q)$ for the time interval $[0, \varepsilon)$:

$$q(\varepsilon) = q_0 + \varepsilon g_1(q_0) + \frac{1}{2}\varepsilon^2 \frac{\partial g_1}{\partial q}(q_0)g_1(q_0) + O(\varepsilon^3)$$

$$= q_0 + \varepsilon g_1 + \frac{1}{2}\varepsilon^2 \frac{\partial g_1}{\partial q}g_1 + O(\varepsilon^3).$$

Step 2. $\dot{q} = g_2(q)$ for the time interval $[\varepsilon, 2\varepsilon)$:

$$q(2\varepsilon) = q(\varepsilon) + \varepsilon g_2(q(\varepsilon)) + \frac{1}{2}\varepsilon^2 \frac{\partial g_2}{\partial q}(q(\varepsilon))g_2(q(\varepsilon)) + O(\varepsilon^3).$$

But from Step 1, we then have

$$g_2(q(\varepsilon)) = g_2(q_0 + \varepsilon g_1(q_0) + O(\varepsilon^2)) = g_2 + \varepsilon \frac{\partial g_2}{\partial q}g_1 + O(\varepsilon^2)$$

and

$$\frac{\partial g_2}{\partial q}(q(\varepsilon))g_2(q(\varepsilon)) = \frac{\partial g_2}{\partial q}g_2 + O(\varepsilon).$$

Thus

$$q(2\varepsilon) = \left\{ q_0 + \varepsilon g_1 + \frac{1}{2}\varepsilon^2 \frac{\partial g_1}{\partial q}g_1 \right\} + \varepsilon \left\{ g_2 + \varepsilon \frac{\partial g_2}{\partial q}g_1 \right\} + \frac{1}{2}\varepsilon^2 \frac{\partial g_2}{\partial q}g_2 + O(\varepsilon^3)$$

$$= q_0 + \varepsilon\{g_1 + g_2\} + \frac{1}{2}\varepsilon^2 \left\{ \frac{\partial g_1}{\partial q}g_1 + 2\frac{\partial g_2}{\partial q}g_1 + \frac{\partial g_2}{\partial q}g_2 \right\} + O(\varepsilon^2).$$

Step 3. $\dot{q} = -g_1(q)$ for the time interval $[2\varepsilon, 3\varepsilon)$:

$$q(3\varepsilon) = q(2\varepsilon) - \varepsilon g_1(q(2\varepsilon)) + \frac{1}{2}\varepsilon^2 \frac{\partial g_1}{\partial q}(q(2\varepsilon))g_1(q(2\varepsilon)) + O(\varepsilon^3).$$

But from Step 2, we then have

$$g_1(q(2\varepsilon)) = g_1(q_0 + \varepsilon(g_1 + g_2) + O(\varepsilon^2)) = g_1 + \varepsilon \frac{\partial g_1}{\partial q}(g_1 + g_2) + O(\varepsilon^2)$$

and

$$\frac{\partial g_1}{\partial q}(q(2\varepsilon))g_1(q(2\varepsilon)) = \frac{\partial g_1}{\partial q}g_1 + O(\varepsilon).$$

After cancelling terms

$$q(3\varepsilon) = q_0 + \varepsilon g_2 + \frac{1}{2}\varepsilon^2 \left\{ \frac{\partial g_2}{\partial q}g_2 + 2\frac{\partial g_2}{\partial q}g_1 - 2\frac{\partial g_1}{\partial q}g_2 \right\} + O(\varepsilon^3).$$

Step 4. $\dot{q} = -g_2(q)$ for the time interval $[3\varepsilon, 4\varepsilon)$:

$$q(4\varepsilon) = q(3\varepsilon) - \varepsilon g_2(q(3\varepsilon)) + \frac{1}{2}\varepsilon^2 \frac{\partial g_2}{\partial q}(q(3\varepsilon))g_2(q(3\varepsilon)) + O(\varepsilon^3).$$

But from Step 3, we then have

$$g_2(q(3\varepsilon)) = g_2(g_0 + \varepsilon g_2 + O(\varepsilon^2)) = g_2 + \varepsilon \frac{\partial g_2}{\partial q} g_2 + O(\varepsilon^2)$$

and

$$\frac{\partial g_2}{\partial q}(q(3\varepsilon))g_2(q(3\varepsilon)) = \frac{\partial g_2}{\partial q} g_2 + O(\varepsilon).$$

After cancelling terms

$$q(4\varepsilon) = q_0 + \varepsilon^2 \left\{ \frac{\partial g_2}{\partial q} g_1 - \frac{\partial g_1}{\partial q} g_2 \right\} + O(\varepsilon^3) = q_0 + \varepsilon^2 [g_1, g_2] + O(\varepsilon^3).$$

In conclusion, starting at q_0, we can instantaneously drive the state q not just in the directions $g_1(q_0)$, $g_2(q_0)$, but also in the direction $[g_1, g_2](q_0)$. Likewise also in the direction $[g_1, [g_1, g_2]](q_0)$, etc. Indeed, in any direction in $\text{Lie}_{q_0}(\{g_1, g_2\})$. Therefore, intuitively, the system

$$\dot{q} = u_1 g_1(q) + u_2 g_2(q)$$

is controllable at the point q if

$$\dim \text{Lie}_q(\{g_1, g_2\}) = \dim (q).$$

The mathematical statement of this fact is known as Chow's Theorem (1939). The unicycle is controllable: e.g., it can be parked in any configuration.

Next we turn to the stabilizability problem of a nonlinear system. For a general nonlinear system

$$\dot{x} = f(x, u),$$

we say it is *stabilizable* by a smooth (or continuous), time-invariant feedback control $u = k(x)$ if the closed-loop system $\dot{x} = f(x, k(x))$ is asymptotically stable. We shall show that the unicycle cannot be stabilizable by a smooth (or even continuous), memoryless state feedback.

Let us first linearize the unicycle model about the origin:

$$\begin{cases} \dot{x} = v \\ \dot{y} = 0 \\ \dot{\theta} = \omega. \end{cases}$$

Clearly, this is not stabilizable. So stabilizing the unicycle can not be done merely by linearization about the equilibrium.

Now we are ready to introduce the Brockett's Theorem that is a necessary condition for being able to stabilize

$$\dot{x} = f(x, u)$$

by a smooth, time-invariant state-feedback $u = k(x)$.

Recall that for a linear control system

$$\dot{x} = Ax + Bu, \quad u = Fx$$
$$\dot{x} = (A + BF)x.$$

with an equilibrium point at the origin, it is asymptotically stable if and only if

$$(\forall\ x(0))\ \lim_{t \to \infty} x(t) = 0.$$

Equivalently, all the eigenvalues of $A + BF$ satisfy $\mathrm{Re}(\lambda) < 0$. A necessary condition for asymptotic stability:

$$0 \text{ is not an eigenvalue of } A + BF$$
$$\Leftrightarrow \mathrm{rank}(A + BF) = n$$
$$\Leftrightarrow \text{the mapping } A + BF \text{ is onto.}$$

Since

$$A + BF = \begin{bmatrix} A\ B \end{bmatrix} \begin{bmatrix} I \\ F \end{bmatrix},$$

a necessary condition for stabilizability (existence of F) is that the mapping

$$f : (x, u) \mapsto Ax + Bu$$

is onto.

Back to the unicycle:

$$\dot{x} = v\cos(\theta)$$
$$\dot{y} = v\sin(\theta)$$
$$\dot{\theta} = \omega.$$

So

$$f : (x, y, \theta, v, \omega) \mapsto (v\cos(\theta), v\sin(\theta), \omega).$$

The mapping $f : \mathbb{R}^5 \to \mathbb{R}^3$ does not satisfy the following condition:

Brockett's condition—The image of every open neighborhood of the origin contains an open neighborhood of the origin.

To see this, it suffices to show that the mapping

$$\begin{bmatrix} \theta \\ v \end{bmatrix} \mapsto \begin{bmatrix} v\cos(\theta) \\ v\sin(\theta) \end{bmatrix} : \mathbb{R}^2 \to \mathbb{R}^2$$

does not satisfy the Brockett's condition. Notice that this function maps the open set

$$|\theta| < \pi/4, \ |v| < 1$$

onto the "bowtie" set (see Fig. 8.6), which obviously does not contain an open

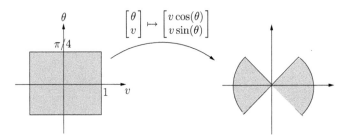

Fig. 8.6. Maps an open set to a "bowtie" set.

neighborhood of the origin.

Intuitively, Brockett's condition says that near the equilibrium $x = 0, u = 0$ there is no restriction on the assignable vector field, that is, one can go in any direction. Now to Brockett's Theorem. The plant $\dot{x} = f(x, u)$ with $f(0, 0) = 0$. The allowable control law is $u = k(x)$ with $k(0) = 0$. Then the controlled system is

$$\dot{x} = f(x, k(x)) =: F(x).$$

If f and k are assumed only continuous, then existence and uniqueness of solutions are not immediate. So like Brockett, we assume f and k are smooth (or at least continuously differentiable), though the theorem has been generalized since its original version.

Theorem 8.1 ([27], Brockett, 1983). *A necessary condition for there to exist a smooth feedback control $u = k(x)$ so that the origin of $\dot{x} = F(x)$ is (locally) asymptotically stable is that f satisfies the Brockett's condition.*

8.3 Information Flow and Allowable Control

Consider a group of n unicycles (Fig. 8.7) in the plane. Let $z_i = (x_i, y_i) \in \mathbb{R}^2$

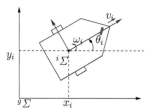

Fig. 8.7. Collective vehicles.

denote the position of robot i, let $\theta_i \in \mathbb{R}$ denote the orientation of robot i, and let v_i, ω_i denote the forward and angular speeds of robot i, respectively. Thus, the group of n unicycles are described by

$$\begin{cases} \dot{x}_i = v_i \cos(\theta_i) \\ \dot{y}_i = v_i \sin(\theta_i) \qquad i = 1, \dots, n. \\ \dot{\theta}_i = \omega_i \end{cases} \tag{8.2}$$

Identify the real plane, $\mathbb{R}^2$, and the complex plane, $\mathbb{C}$, by identifying a vector, z_i, and a complex number, $\mathbf{z}_i$. The kinematic model in complex form is

$$\begin{cases} \dot{\mathbf{z}}_i = v_i e^{j\theta_i} \\ \dot{\theta}_i = \omega_i \end{cases} \qquad i = 1, \dots, n. \tag{8.3}$$

Following [64], we construct a moving frame $^i\varSigma$, the Frenet-Serret frame, that is fixed on the vehicle (see Fig. 8.8). Let r_i be the unit vector tangent to the

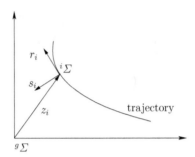

Fig. 8.8. Frenet-Serret frame.

trajectory in the direction of motion at the current location of the vehicle (r_i is the normalized velocity vector) and let s_i be r_i rotated by $\pi/2$. Since the vehicle is moving at speed v_i, $v_i r_i = \dot{z}_i$, and so in complex form

$$\mathbf{r}_i = e^{j\theta_i}, \quad \mathbf{s}_i = j\mathbf{r}_i.$$

Thus

$$\dot{\mathbf{r}}_i = \frac{d}{dt}\left(e^{j\theta_i}\right) = je^{j\theta_i}\dot{\theta}_i = \mathbf{s}_i\omega_i, \quad \dot{\mathbf{s}}_i = j\dot{\mathbf{r}}_i = j\mathbf{s}_i\omega_i = -\mathbf{r}_i\omega_i.$$

The kinematic equations using the Frenet-Serret frame are therefore

$$\begin{cases} \dot{z}_i = v_i r_i \\ \dot{r}_i = \omega_i s_i \\ \dot{s}_i = -\omega_i r_i \end{cases} \quad i = 1,\dots,n. \tag{8.4}$$

Now we define the information flow graph for the group of n unicycles as we did for the point-mass robots. We refer to individual unicycles as nodes and information flows as links.

Definition 8.1. *The* information flow graph $\mathcal{G}$ *consists of*

- *a node set* $\mathcal{V} = \{1, 2\dots, n\}$, *each node i corresponding to unicycle i;*
- *a set $\mathcal{E}$ of arcs: The arc from node j to node i is one of its arcs just in case unicycle i can access the information from unicycle j in some way.*

By this definition, information flows in the direction of the arcs in $\mathcal{G}$. Let $\mathcal{N}_i$ denote the set of labels of those unicycles whose information flows to unicycle i. Throughout this chapter, we have the following assumption.

Assumption: The information flow graph $\mathcal{G}$ is fixed.

Fixed graph means that the information flow topology is static and so is the neighbor set $\mathcal{N}_i$. Although modeling the information flow with a static graph may not accurately model realistic situations whereby sensors have a limited field of view, it is a necessary step toward the more realistic dynamic setting where ad-hoc links can be established or dropped.

In our setup, no unicycle can access the absolute positions of other unicycles or its own. Unicycle i can measure only the relative positions of its neighbor

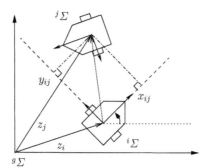

Fig. 8.9. Local information.

unicycles defined in the information flow graph with respect to its own Frenet-Serret frame (see Fig. 8.9). Suppose unicycle can sense unicycle j. Write the relative position $z_j - z_i$ in i's local frame:

$$\begin{cases} x_{ij} = (z_j - z_i) \cdot r_i \\ y_{ij} = (z_j - z_i) \cdot s_i \end{cases} \quad j \in \mathcal{N}_i, \tag{8.5}$$

where dot denotes dot product. Thus, we assume unicycle i can obtain x_{ij} and y_{ij} about unicycle j, which will be used as feedback.

Our objective is to study how the information flow structure affects the feasibility of stabilization of n unicycles to desired formations. For a formation in the plane, we emphasize that only the positions of the group of n unicycles are required and there is no constraints on their orientations. Let $z = (z_1, \ldots, z_n) \in \mathbb{R}^{2n}$ denote the aggregate position state and let $\theta = (\theta_1, \ldots, \theta_n) \in \mathbb{R}^n$ denote the aggregate orientation state of n unicycles. Hence, an *allowable controller* looks like

$$\begin{cases} v_i = g_i\Big(t, (x_{ij}, y_{ij})|_{j \in \mathcal{N}_i}\Big) \\ \omega_i = h_i\Big(t, (x_{ij}, y_{ij})|_{j \in \mathcal{N}_i}\Big) \end{cases} \qquad i = 1, \ldots, n \qquad (8.6)$$

where g_i, h_i are smooth functions of their arguments. Since we showed in the previous section that the unicycle is controllable but is not stabilizable by a smooth state feedback due to Brockett's condition, we allow the distributed control for the formation problem to be time-varying (i.e., an explicit variable t in (8.6)) though it has not been argued in the literature whether the group of unicycles are able to be stabilized to a formation by a smooth state feedback.

8.4 Rendezvous

In this section we study the rendezvous problem of n unicycles with a fixed information flow graph $\mathcal{G}$. We emphasize here that only the positions of the group of n unicycles are required to agree on certain common location for the rendezvous problem and their orientations are not required. That is,

Definition 8.2. *A group of n unicycles (8.2) rendezvous if for any initial state $z_i(0), \theta_i(0), i = 1, \ldots, n$, there exists a point $\zeta \in \mathbb{R}^2$ such that for all i, $z_i(t)$ converges to ζ.*

Let us first write out a bit results for $n > 1$ point-mass robots modeled by complex numbers, $z_1, \ldots, z_n$, in the plane. Each robot has velocity control:

$$\dot{z}_i = u_i.$$

Each robot senses the relative positions of a subgroup, $\mathcal{N}_i$, of the others. Let ξ_i denote the vector with components $z_j - z_i$, $j \in \mathcal{N}_i$. Thus ξ_i represents the information available to u_i. Controllers are of the form

$$u_i = F_i \xi_i.$$

For fixed information flow graph, the rendezvous problem is to find, if possible, $F_1, \ldots, F_n$ such that

$$\forall \, z(0))(\exists \, \zeta \in \mathbb{C}) \, \lim_{t \to \infty} z(t) = \zeta \mathbf{1}.$$

And a solution to the problem is the following theorem that can be easily obtained from Theorem 5.3.

Theorem 8.2. 1. *Rendezvous is feasible for a group of point-mass robots if and only if the information flow graph is quasi-strongly connected.*
2. *A feasible control to solve the rendezvous problem is*

$$F_i = \begin{bmatrix} 1 \cdots 1 \end{bmatrix}.$$

The goal in the following is to extend this result to unicycles.

Theorem 8.3. *Rendezvous is feasible for a group of n unicycles (8.2) if and only if the information flow graph $\mathcal{G}$ is quasi-strongly connected.*

Proof: ($\Longrightarrow$) Suppose the group of n unicycles rendezvous with the control (8.6). Thus, we obtain the closed-loop system for $i = 1, \ldots, n$,

$$\begin{cases} \dot{z}_i = g_i\Big(t, (x_{ij}, y_{ij})|_{j \in \mathcal{N}_i}\Big) r_i = g_i\Big(t, \big[(z_j - z_i) \cdot r_i, \ (z_j - z_i) \cdot s_i\big]|_{j \in \mathcal{N}_i}\Big) r_i \\ \dot{\theta}_i = h_i\Big(t, (x_{ij}, y_{ij})|_{j \in \mathcal{N}_i}\Big) = h_i\Big(t, \big[(z_j - z_i) \cdot r_i, \ (z_j - z_i) \cdot s_i\big]|_{j \in \mathcal{N}_i}\Big) \end{cases} \tag{8.7}$$

and we have that for a given initial condition z_i^0 and θ_i^0 ($i = 1, \ldots, n$), there exists $\zeta \in \mathbb{R}^2$ such that

$$z_i(t) \to \zeta \text{ for all } i \text{ as } t \to \infty.$$

By way of contradiction, suppose that the information flow graph $\mathcal{G}$ is not quasi-strongly connected. Then by the definition of quasi-strong connectivity, it follows that there are two nodes i^* and j^* such that for any node k, either i^* or j^* is not reachable from k. Let $\mathcal{V}_1$ be the subset of nodes from which i^* is reachable and $\mathcal{V}_2$ be the subset of nodes from which j^* is reachable. By this construction, we know that $\mathcal{V}_1$ and $\mathcal{V}_2$ are disjoint. Moreover, for each node $i \in \mathcal{V}_1$ (resp. $\mathcal{V}_2$), its neighbor set $\mathcal{N}_i$ also belongs to $\mathcal{V}_1$ (resp. $\mathcal{V}_2$). It means the dynamics of the subgroup of unicycles in $\mathcal{V}_2$ are not affected by the states of unicycles in $\mathcal{V}_1$. In other words, the trajectories of the subgroup of unicycles in $\mathcal{V}_2$ are independent of the choice of the initial states of unicycles in $\mathcal{V}_1$. So as long as $z_i(0) = z_i^0$ and

$\theta_i(0) = \theta_i^0$ for all $i \in \mathcal{V} \setminus \mathcal{V}_1$, the limit point of z_i, $i \in \mathcal{V} \setminus \mathcal{V}_1$ does not change, namely, still ζ. Also, since the n unicycles rendezvous, we then have

$$z_i(t) \to \zeta \text{ for all } i \text{ as } t \to \infty.$$

Now let $a \in \mathbb{R}^2$ satisfying $a \neq 0$. For $i \in \mathcal{V}_1$, we apply the coordinate transformation:

$$z_i' = z_i + a.$$

Since $\mathcal{N}_i \subseteq \mathcal{V}_1$ for all $i \in \mathcal{V}_1$, we obtain

$$\begin{cases} \dot{z}_i' = g_i\left(t, \left[(z_j' - z_i') \cdot r_i, \ (z_j' - z_i') \cdot s_i\right]\big|_{j \in \mathcal{N}_i}\right) r_i \\ \dot{\theta}_i = h_i\left(t, \left[(z_j' - z_i') \cdot r_i, \ (z_j' - z_i') \cdot s_i\right]\big|_{j \in \mathcal{N}_i}\right) \end{cases} \text{ for } i \in \mathcal{V}_1.$$

Notice that the above dynamics (after coordinate transformation) for the subgroup of unicycles in $\mathcal{V}_1$ is the same as the dynamics (8.7) for the same subgroup of unicycles. Hence, with the initial condition

$$z_i'(0) = z_i^0, \ \theta_i(0) = \theta_i^0 \text{ for } i \in \mathcal{V}_1$$
$$z_j(0) = z_j^0, \ \theta_j(0) = \theta_j^0 \text{ for } j \in \mathcal{V} \setminus \mathcal{V}_1$$

we can conclude that

$$z_i'(t) \to \zeta \text{ for all } i \in \mathcal{V}_1 \text{ as } t \to \infty.$$

By recalling that $z_i' = z_i + a$, this in turn implies that, when

$$z_i(0) = z_i^0 - a, \ \theta_i(0) = \theta_i^0 \text{ for } i \in \mathcal{V}_1$$
$$z_j(0) = z_j^0, \ \theta_j(0) = \theta_j^0 \text{ for } j \in \mathcal{V} \setminus \mathcal{V}_1$$

the trajectory

$$z_i(t) \to (\zeta - a) \text{ for } i \in \mathcal{V}_1,$$

a contradiction.

($\Longleftarrow$) The sufficiency of the proof is left to Theorem 8.4.   ∎

In order for unicycles to rendezvous, we now present an explicit smooth periodic feedback controller, which is inspired and modified from the time-varying control law in [147, 148].

Theorem 8.4. *Assume the information flow graph $\mathcal{G}$ is quasi-strongly connected. Then the n unicycles (8.2) rendezvous with the distributed control*

$$\begin{cases} v_i = k \sum_{j \in \mathcal{N}_i} a_{ij} x_{ij} \\ \omega_i = \cos(t) \end{cases} \tag{8.8}$$

where $k > 0$ sufficiently small and $a_{ij} > 0$.

Proof: Let $a_{ij} = 0$ if $j \notin \mathcal{N}_i$. Then v_i in (8.8) can be rewritten as

$$v_i = k \sum_{j \neq i} a_{ij} x_{ij} = k \sum_{j \neq i} a_{ij} \left(z_j - z_i \right) \cdot r_i.$$

Next, using the identity

$$(z_i \cdot r_i) r_i = (r_i r_i^T) z_i,$$

we obtain

$$\dot{z}_i = v_i r_i = k \sum_{j \neq i} a_{ij} \left((z_j - z_i) \cdot r_i \right) r_i = k \, (r_i r_i^T) \sum_{j \neq i} a_{ij} \left(z_j - z_i \right).$$

Define

$$M\left(\theta_i(t)\right) := r_i r_i^T = \begin{pmatrix} \cos^2(\theta_i(t)) & \cos(\theta_i(t))\sin(\theta_i(t)) \\ \cos(\theta_i(t))\sin(\theta_i(t)) & \sin^2(\theta_i(t)) \end{pmatrix}$$

and

$$H(\theta(t)) := \begin{pmatrix} M(\theta_1(t)) & 0 & 0 & 0 \\ 0 & M(\theta_2(t)) & 0 & 0 \\ 0 & 0 & \ddots & 0 \\ 0 & 0 & 0 & M(\theta_n(t)) \end{pmatrix}. \tag{8.9}$$

Thus, the overall position dynamics read as

$$\dot{z} = kH(\theta(t))\left(A \otimes I_2\right)z, \tag{8.10}$$

where $z \in \mathbb{R}^{2n}$ is the aggregate state of $z_1, z_2, \ldots, z_n$ and A is an $n \times n$ matrix whose off-diagonal entries are a_{ij} and diagonal entries are $\sum_{j \neq i} a_{ij}$. Hence, the matrix A is a generator matrix.

Let

$$\bar{M}_i = \frac{1}{2\pi} \int_0^{2\pi} M\left(\theta_i(\tau)\right) d\tau.$$

Thus we obtain

$$\bar{M}_i := \begin{pmatrix} m_i^1 & m_i^2 \\ m_i^2 & m_i^3 \end{pmatrix}$$

where

$$m_i^1 = \frac{1}{2\pi} \int_0^{2\pi} \cos^2(\theta_i(\tau)) d\tau,$$

$$m_i^2 = \frac{1}{2\pi} \int_0^{2\pi} \cos(\theta_i(\tau)) \sin(\theta_i(\tau)) d\tau,$$

$$m_i^3 = \frac{1}{2\pi} \int_0^{2\pi} \sin^2(\theta_i(\tau)) d\tau.$$

Let $H_{av} = \text{diag}(\bar{M}_1, \ldots, \bar{M}_n)$. Then the averaged system associated with the system (8.10) is given as follows

$$\dot{z} = H_{av}(A \otimes I_2) z. \tag{8.11}$$

Applying the Cauchy-Schwarz inequality leads to

$$m_i^1 m_i^3 \geq (m_i^2)^2.$$

Furthermore, since $\theta_i(t)$ is not constant, the inequality holds strictly. It now follows that $\bar{M}_i$ is positive definite and therefore H_{av} is positive definite. More exactly, we say H_{av} is α-diagonal positive definite with

$$\alpha = \big\{ \{1, 2\}, \{3, 4\}, \ldots, \{2n - 1, 2n\} \big\}.$$

Recall that $\mathcal{G}$ is an opposite digraph of $\mathcal{G}_A$, the associated digraph of the generator matrix A. If $\mathcal{G}$ is quasi-strongly connected then $\mathcal{G}_A$ has a globally reachable node by Theorem 2.1 and the matrix $A \otimes I_2$ is $H(\alpha, 2)$ stable by Theorem 4.6. This means that there is a similarity transformation P such that

$$P^{-1}(A \otimes I_2) P = \begin{pmatrix} A_s & 0 \\ 0 & 0_{2 \times 2} \end{pmatrix},$$

where A_s is Hurwitz and the last two column vectors of F must be in the null space of $A \otimes I_2$. Without loss of generality, we choose the last two column vectors of P to be $\mathbf{1} \otimes I_{2 \times 2}$.

Applying the transformation

$$e = \begin{pmatrix} e_s \\ e_o \end{pmatrix} = P^{-1}z$$

where $e_s \in \mathbb{R}^{2n-2}$ and $e_o \in \mathbb{R}^2$, to the system (8.10) yields

$$\begin{pmatrix} \dot{e}_s \\ \dot{e}_o \end{pmatrix} = kP^{-1}H\left(\theta(t)\right)\left(A \otimes I_2\right)Pe$$

$$= kP^{-1}H\left(\theta(t)\right)P \begin{pmatrix} A_s & 0 \\ 0 & 0_{2\times 2} \end{pmatrix} \begin{pmatrix} e_s \\ e_s \end{pmatrix}$$

$$=: k \begin{pmatrix} \tilde{A}_s(t) & 0 \\ \tilde{B}(t) & 0_{2\times 2} \end{pmatrix} \begin{pmatrix} e_s \\ e_o \end{pmatrix}.$$

Correspondingly, for the averaged system (8.11), we apply the same coordinate transformation. Then we have

$$\begin{pmatrix} \dot{e}_s \\ \dot{e}_o \end{pmatrix} = P^{-1}H_{av}\left(A \otimes I_2\right)Pe$$

$$= P^{-1}H_{av}P \begin{pmatrix} A_s & 0 \\ 0 & 0_{2\times 2} \end{pmatrix} \begin{pmatrix} e_s \\ e_o \end{pmatrix}$$

$$=: \begin{pmatrix} \bar{A}_s & 0 \\ \bar{B} & 0_{2\times 2} \end{pmatrix} \begin{pmatrix} e_s \\ e_o \end{pmatrix}.$$

Since $A \otimes I_2$ is $H(\alpha, 2)$ stable, the matrix $\bar{A}_s$ has to be Hurwitz. Hence the reduced averaged system

$$\dot{e}_s = \bar{A}_s e_s$$

is exponentially stable. Then, it follows from Theorem D.1 (Averaging Theorem) that there exists a positive constant k^* such that, for all $0 < k < k^*$, global exponential stability of the reduced original system

$$\dot{e}_s = k\tilde{A}_s(t)e_s$$

is established.

Moreover, since $\dot{e}_o = k\tilde{B}(t)e_s$ and $\tilde{B}(t)$ is uniformly bounded, it follows that $\dot{e}_o \to 0$ exponentially as $t \to \infty$. This implies that e_o tends to some finite constant vector, say $\zeta \in \mathbb{R}^2$. In conclusion,

$$\lim_{t\to\infty} z(t) = \lim_{t\to\infty} P \begin{pmatrix} e_s(t) \\ e_o(t) \end{pmatrix} = \mathbf{1} \otimes \zeta.$$

Thus, we have proven that the n unicycles rendezvous with the distributed control (8.8). ∎

An alternative choice of distributed control to make the n unicycles rendezvous is

$$\begin{cases} v_i(t) = \sum_{j \in \mathcal{N}_i} a_{ij} x_{ij}(t) \\ \omega_i(t) = \gamma \cos(\gamma t) \end{cases} \quad i = 1, 2, \ldots, n \quad (8.12)$$

where $a_{ij} > 0$ and $\gamma > 0$ sufficiently large. By applying a time scaling, $\tau = \frac{t}{\gamma}$, one can use the same argument and know that there exists a positive constant γ^* such that, for all $\gamma^* < \gamma < \infty$, the n vehicles rendezvous with the control (8.12).

The above results show that the n vehicles can be made exponentially converge to a common location through simple local actions, or we say they achieve an agreement about a common point if the information flow graph is quasi-strongly connected. Notice that this graphical condition is equivalent to that $\mathcal{G}$ has a center node. So if we treat a beacon placed at the proper location as one member of the group of vehicles and it is the center node in the information flow graph, the local actions of each individual vehicle result in the group gathering at the beacon.

Consider ten unicycles in the plane. The information flow digraph is given in Fig. 8.10. Clearly, it is quasi-strongly connected. For example, the node 5 is a center node in the graph, while the node 1 is not. The ten unicycles are randomly initialized and the controller (8.8) is used with the choice of $k = 1$. A simulation trajectories are presented in Fig. 8.11, which shows that they eventually come to a same location (rendezvous).

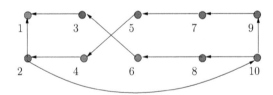

Fig. 8.10. Information flow digraph.

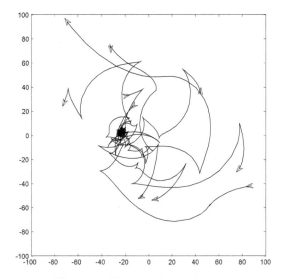

Fig. 8.11. Ten unicycles rendezvous.

8.5 Row straightening

In this section we investigate line formation behaviors of collective unicycles. The group of unicycles move according to a distributed control so that they eventually converge to form a line. Again, only the positions of unicycles are required without any concern on their orientations. First, we would like to find out what is the least restrictive condition on the information flow graph $\mathcal{G}$ such that a line formation is feasible for a group of n unicycles.

Let $\mathcal{G}^*$ denote the opposite digraph of the information flow graph $\mathcal{G}$. (The definition of opposite digraph can be found in Section 2.4.)

Theorem 8.5. *Line formation is feasible for a group of n unicycles (8.2) if and only if $\mathcal{G}^*$ has at most two closed strong components.*

Proof: ($\Longleftarrow$) Firstly, suppose that $\mathcal{G}^*$ has one closed strong component (see Fig. 8.12). Then by Theorem 2.1 $\mathcal{G}$ is quasi-strongly connected and so by Theorem 8.4 the n unicycles are able to rendezvous (come to a common point), which can be treated as a trivial case of line formation.

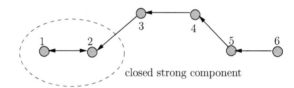

closed strong component

Fig. 8.12. $\mathcal{G}^*$ has one closed strong component.

Secondly, suppose that $\mathcal{G}^*$ has two closed strong components, say $\mathcal{G}_1^* = (\mathcal{V}_1, \mathcal{E}_1^*)$, $\mathcal{G}_2^* = (\mathcal{V}_2, \mathcal{E}_2^*)$, and in addition that $\mathcal{V}_1 \cup \mathcal{V}_2 = \mathcal{V}$ (see Fig. 8.13).

Then both $\mathcal{G}_1$ and $\mathcal{G}_2$ are quasi-strongly connected by Theorem 2.1. Thus, it follows from Theorem 8.4 that each of the corresponding subgroup of unicycles are able to rendezvous. Again, it is a trivial line formation for the whole group. Thirdly, suppose that $\mathcal{G}^*$ has two closed strong components, say $\mathcal{G}_1^* = (\mathcal{V}_1, \mathcal{E}_1^*)$,

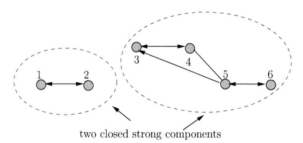

two closed strong components

Fig. 8.13. $\mathcal{G}^*$ has two closed strong components but $\mathcal{V} = \mathcal{V}_1 \cup \mathcal{V}_2$.

$\mathcal{G}_2^* = (\mathcal{V}_2, \mathcal{E}_2^*)$, but $\mathcal{V}_1 \cup \mathcal{V}_2 \neq \mathcal{V}$ (see Fig. 8.14). Let $\mathcal{V}_3 = \mathcal{V} - \mathcal{V}_1 - \mathcal{V}_2$. Without

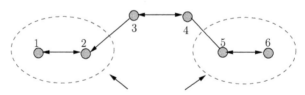

two closed strong components

Fig. 8.14. $\mathcal{G}^*$ has two closed strong components but $\mathcal{V} \neq \mathcal{V}_1 \cup \mathcal{V}_2$.

loss of generality, assume

$$\mathcal{V}_1 = \{1, \ldots, r_1\}, \ \mathcal{V}_2 = \{r_1 + 1, \ldots, r_1 + r_2\}, \ \text{and} \ \mathcal{V}_3 = \{r_1 + r_2 + 1, \ldots, n\}.$$

Let $r_3 = n - r_1 - r_2$. Consider the time-varying feedback controller (8.8) again. Then the closed-loop overall system is given as follows:

$$\dot{z} = kH(\theta(t))(A \otimes I_2) z \qquad (8.13)$$

where $H(\theta(t))$ is the same as (8.9) and A is a generator matrix. Further, the matrix A has the form

$$A = \begin{pmatrix} A_1 & 0 & 0 \\ 0 & A_2 & 0 \\ B_1 & B_2 & A_3 \end{pmatrix},$$

where $A_1 \in \mathbb{R}^{r_1 \times r_1}$, $A_2 \in \mathbb{R}^{r_2 \times r_2}$, and $A_3 \in \mathbb{R}^{r_3 \times r_3}$. Moreover, one can easily verify that $-A_3$ is a nonsingular M-matrix by Theorem 3.6. Thus the system (8.13) can be written as

$$\begin{cases} \dot{z}^1 = kH^1(\theta(t))(A_1 \otimes I_2) z^1, \\ \dot{z}^2 = kH^2(\theta(t))(A_2 \otimes I_2) z^2, \\ \dot{z}^3 = kH^3(\theta(t))\Big((A_3 \otimes I_2)z^3 + (B_1 \otimes I_2)z^1 + (B_2 \otimes I_2)z^2\Big), \end{cases}$$

where z^i, $i = 1, 2, 3$ are of compatible dimensions and $H^1(\theta(t)), H^2(\theta(t)), H^3(\theta(t))$ are suitable matrices.

Since the associated digraph $\mathcal{G}_{A_1}$ is exactly the digraph $\mathcal{G}_1^*$, which is strongly connected, then by the same argument as in the proof of Theorem 8.4, there exists a positive constant k_1^* such that, for all $0 < k < k_1^*$,

$$\lim_{t \to \infty} z^1(t) = \mathbf{1} \otimes \zeta_1 \quad \text{for some } \zeta_1 \in \mathbb{R}^2.$$

For the same reason, there exists k_2^* such that, for all $0 < k < k_2^*$,

$$\lim_{t \to \infty} z^2(t) = \mathbf{1} \otimes \zeta_2 \quad \text{for some } \zeta_2 \in \mathbb{R}^2.$$

Moreover, they have exponential convergence rate.

Next applying the change of variables

$$\varsigma = (A_3 \otimes I_2)z^3 + (B_1 \otimes I_2)z^1 + (B_2 \otimes I_2)z^2$$

yields

$$\varsigma = k(A_3 \otimes I_2)H^3(\theta(t))\varsigma + k(B_1 \otimes I_2)H^1(\theta(t))(A_1 \otimes I_2)z^1$$
$$+ k(B_2 \otimes I_2)H^2(\theta(t))(A_2 \otimes I_2)z^2. \tag{8.14}$$

Since $-A_3$ is a nonsingular M-matrix, and so is $-A_3^T$. It follows from the same argument in the proof of Theorem 4.6 that $(A_3 \otimes I_2)^T$ is $H(\alpha, 0)$ stable where

$$\alpha = \big\{\{1, 2\}, \ldots, \{r_3 - 1, r_3\}\big\}.$$

This implies that, for the α block diagonal symmetric positive matrix $\bar{H}^3$, the average of $H^3(\theta(t))$, the matrix $\bar{H}^3 (A_3 \otimes I_2)^T$ is Hurwitz and so is $(A_3 \otimes I_2)\bar{H}^3$. Invoking Theorem D.1 (Averaging theorem) here gives that, there exists a positive constant k_3^* such that, for all $0 < k < k_3^*$, the origin of the nominal system

$$\varsigma = k(A_3 \otimes I_2)H^3(\theta(t))\varsigma$$

is globally exponentially stable.

In addition, notice that the other two terms in (8.14) both exponentially converge to zero. Hence (8.14) can be viewed as an exponentially stable system with an exponentially vanishing input, and so its origin is exponentially stable.

Let

$$k^* = \min\{k_1^*, k_2^*, k_3^*\}.$$

Thus, for all $0 < k < k^*$,

$$\lim_{t \to \infty} z^3(t) = -(A_3 \otimes I_2)^{-1}(B_1 \otimes I_2)\lim_{t \to \infty} z^1(t) - (A_3 \otimes I_2)^{-1}(B_2 \otimes I_2)\lim_{t \to \infty} z^2(t)$$
$$= -(A_3 \otimes I_2)^{-1}(B_1 \otimes I_2)(\mathbf{1} \otimes \zeta_1) - (A_3 \otimes I_2)^{-1}(B_2 \otimes I_2)(\mathbf{1} \otimes \zeta_2)$$
$$= -\left(A_3^{-1}B_1\mathbf{1}\right) \otimes \zeta_1 - \left(A_3^{-1}B_2\mathbf{1}\right) \otimes \zeta_2.$$

Since

$$\big(B_1 \ B_2 \ A_3\big)\mathbf{1} = 0,$$

we have

$$-\left(A_3^{-1}B_1\mathbf{1}\right) - \left(A_3^{-1}B_2\mathbf{1}\right) = \mathbf{1}.$$

Hence, all $z_i(t)$ for $i \in \mathcal{V}_3$ approach a convex combination of ζ_1 and ζ_2, which implies that a line formation is achieved.

($\Longrightarrow$) The necessary proof follows from the same idea as the one for Theorem 8.4. In other words, if $\mathcal{G}^*$ has at least three closed strong components (see Fig. 8.15), then by the same argument shown in the proof of Theorem 8.4, we

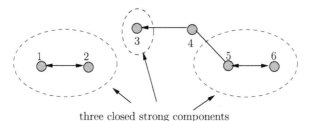

three closed strong components

Fig. 8.15. $\mathcal{G}^*$ has three closed strong components.

can find an initial condition for which the trajectory $z(t)$ converges to form a triangle. Hence, a line formation is infeasible. ∎

We now introduce a special information flow graph which guarantees that all vehicles converge to a line segment, equally spaced. This is an extension to unicycles of a line-formation scheme of Wagner and Bruckstein [139]. Let $\mathcal{G}^*$ have two closed strong components, both of which consisting of only one node, say 1 and n. Vehicles 1 and n are called *edge leaders*. The edge leaders here are not necessarily wheeled vehicles. They can be virtual beacons or landmarks. But the vehicles respond to these edge leaders much like they respond to real neighbor vehicles. The purpose of the edge leaders is to introduce the mission: to direct the vehicle group behavior. We emphasize that the edge leaders are not central coordinators. They do not broadcast instructions. As for the remaining vehicles, i, $i = 2, \ldots, n - 1$, we assume that each vehicle can access the information from vehicles $i - 1$ and $i + 1$. This gives the information flow digraph $\mathcal{G}$ in Fig. 8.16. It is readily seen that the opposite digraph $\mathcal{G}^*$ has exactly two closed

Fig. 8.16. Information flow graph $\mathcal{G}$.

strong components (Fig. 8.17). We now show that in this special case all vehicles

closed strong component

Fig. 8.17. Opposite digraph $\mathcal{G}^*$.

converge to a uniform distribution on the line segment specified by the two edge leaders.

Theorem 8.6. *Consider a group of $n - 2$ robots labelled 2 through $n - 1$ and two stationary edge leaders labelled 1 and n. Then the group of robots converge to a uniform distribution on the line specified by the two edge leaders with the distributed control*

$$
\begin{cases}
v_i(t) = k \sum_{j=\mathcal{N}_i} x_{ij}(t), & \mathcal{N}_i = \{i-1, i+1\} \\
w_i(t) = \cos(t),
\end{cases}
\qquad i = 2, \ldots, n-1 \qquad (8.15)
$$

where $k > 0$ is sufficiently small.

Proof: First we observe that for the controller (8.15) the overall closed-loop system is of the form (8.13), where the matrix A is given as follows:

$$
A =
\begin{pmatrix}
0 & 0 & 0 & 0 & \cdots & 0 \\
1 & -2 & 1 & 0 & \cdots & 0 \\
0 & 1 & -2 & 1 & \cdots & 0 \\
\vdots & \ddots & \ddots & \ddots & \ddots & \vdots \\
0 & \cdots & 0 & 1 & -2 & 1 \\
0 & \cdots & 0 & 0 & 0 & 0
\end{pmatrix}.
$$

Since $\mathcal{G}^*$ has two closed strong components, from the proof of Theorem 8.5 we know that there exists $k^* > 0$ such that, for all $0 < k < k^*$, the n vehicles achieve aline formation. That is,

$$
\lim_{t \to \infty} (A \otimes I_2) z(t) = 0.
$$

Consider the following partition of $\{1, 2, \ldots, n\}$,

$$
\{m_1, m_2, \ldots, m_{2n}\} = \{1, 3, 5, \ldots, 2n - 1, 2, 4, 6, \ldots, 2n\}.
$$

Then the associated permutation matrix P has the unit coordinate vectors

$$\mathbf{e}_{m_1}, \mathbf{e}_{m_2}, \ldots, \mathbf{e}_{m_{2n}}$$

for its columns. Now observe that the matrix P performs the transformation

$$P^T(A \otimes I_2)P = I_2 \otimes A = \begin{pmatrix} A & 0 \\ 0 & A \end{pmatrix} \quad \text{and} \quad P^T z = \begin{pmatrix} x \\ y \end{pmatrix},$$

where $x = (x_1 \cdots x_n)^T$ and $y = (y_1 \cdots y_n)^T$. Thus,

$$0 = (A \otimes I_2)z(\infty) = P(I_2 \otimes A)P^T z(\infty) = P \begin{pmatrix} A & 0 \\ 0 & A \end{pmatrix} \begin{pmatrix} x(\infty) \\ y(\infty) \end{pmatrix}$$

implies

$$Ax(\infty) = 0 \quad \text{and} \quad Ay(\infty) = 0.$$

Also note that

$$\text{Ker}(A) = span \left\{ \xi_1 = \begin{pmatrix} 0 \\ 1 \\ 2 \\ \vdots \\ n-1 \end{pmatrix}, \; \xi_2 = \begin{pmatrix} n-1 \\ n-2 \\ n-3 \\ \vdots \\ 0 \end{pmatrix} \right\}.$$

Hence

$$x(\infty) = \alpha_1 \xi_1 + \alpha_2 \xi_2 \quad \text{and} \quad y(\infty) = \beta_1 \xi_1 + \beta_2 \xi_2.$$

Furthermore, since

$$x_1(\infty) = x_1(0), \; x_n(\infty) = x_n(0) \quad \text{and} \quad y_1(\infty) = y_1(0), \; y_n(\infty) = y_n(0),$$

we solve for

$$\alpha_1 = \frac{x_n(0)}{n-1}, \; \alpha_2 = \frac{x_1(0)}{n-1} \quad \text{and} \quad \beta_1 = \frac{y_n(0)}{n-1}, \; \beta_2 = \frac{y_1(0)}{n-1}.$$

In conclusion, we have

$$\begin{pmatrix} x_1(\infty) \\ x_2(\infty) \\ \vdots \\ x_{n-1}(\infty) \\ x_n(\infty) \end{pmatrix} = \begin{pmatrix} x_1(0) \\ \frac{(n-2)x_1(0)+x_n(0)}{n-1} \\ \vdots \\ \frac{x_1(0)+(n-2)x_n(0)}{n-1} \\ x_n(0) \end{pmatrix}, \quad \begin{pmatrix} y_1(\infty) \\ y_2(\infty) \\ \vdots \\ y_{n-1}(\infty) \\ y_n(\infty) \end{pmatrix} = \begin{pmatrix} y_1(0) \\ \frac{(n-2)y_1(0)+y_n(0)}{n-1} \\ \vdots \\ \frac{y_1(0)+(n-2)y_n(0)}{n-1} \\ y_n(0) \end{pmatrix}.$$

This shows that all vehicles asymptotically approach a uniform distribution on the line with the controller (8.15).   ∎

We make a simulation of six unicycles in the plane where two edge leaders are stationary and the remaining four unicycles moves around the plane using the control law (8.15) with $k = 1$. The information flow digraph is the same as the one in Fig. 8.16. As shown by the simulation in Fig. 8.18, they converge to a line and are distributed in the line segment uniformly.

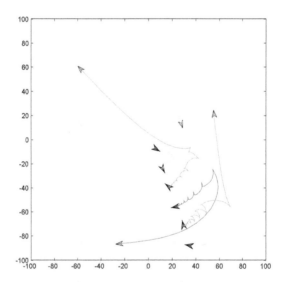

Fig. 8.18. Uniform distribution.

8.6 Pattern Formation

In this section, we turn our attention to the stabilization problem of any pattern formations. Similar to the formation of point-mass robots, we represent a pattern formation of n unicycles in the plane by specifying a vector $c = (c_1, \ldots, c_n)$, where $c_i \in \mathbb{R}^2$. Now define a rotation matrix

$$R(\psi) = \begin{pmatrix} \cos(\psi) & -\sin(\psi) \\ \sin(\psi) & \cos(\psi) \end{pmatrix}$$

and a translation vector

$$\xi \in \mathbb{R}^2.$$

Then $c = (c_1, \ldots, c_n)$ and $c' = (c_1', \ldots, c_n')$, where $c_i' = R(\psi)c_i + \xi$, represent the same patter formation in the plane. In other words, such a formation is up to translation and rotation (see Fig. 8.19 for example).

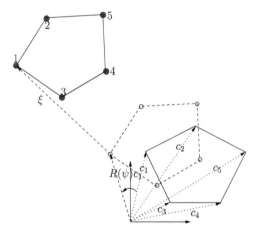

Fig. 8.19. Vehicles in formation.

Definition 8.3. *A group of n robots (8.2) achieve a formation $c = (c_1, \ldots, c_n)$ if for any initial state $z_i(0), \theta_i(0), i = 1, \ldots, n$, there exist a rotation ψ and a translation $\xi \in \mathbb{R}^2$ such that for all i, $z_i(t)$ converges to $R(\psi)c_i + \xi$.*

In order for a group of vehicles to achieve a formation specified by c, suppose that they have a *common sense of direction*, represented by the angle ϑ in Fig. 8.20 and assume that each individual i can measure its own orientation with respect to the common sense of direction, i.e, the angle ϕ_i in Fig. 8.20. This can be implemented. For instance, each vehicle carries a navigation device such as a compass. Alternatively, all vehicles initially agree on their orientation and use it as the common direction. The common direction may not coincide with the positive x-axis of the global frame. Thus $\phi_i = \theta_i - \vartheta$ (see Fig. 8.20).

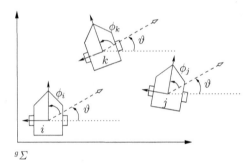

Fig. 8.20. Common sense of direction.

Let us first define a rotation matrix

$$R(\alpha) := \begin{pmatrix} R_1(\alpha) \\ R_2(\alpha) \end{pmatrix} := \begin{pmatrix} \cos(\alpha) & -\sin(\alpha) \\ \sin(\alpha) & \cos(\alpha) \end{pmatrix}.$$

Our result for stabilization of any pattern formation is stated next.

Theorem 8.7. *Assume the information flow graph $\mathcal{G}$ is quasi-strongly connected. Then the n robots (8.2) achieve a formation $c = (c_1, \ldots, c_n)$ with the distributed control*

$$\begin{cases} v_i = \sum_{j \in \mathcal{N}_i} a_{ij}\left(x_{ij} + R_1(-\phi_i)(c_j - c_i)\right) \\ \omega_i = \cos(t) \end{cases} \quad i = 1, \ldots, n \qquad (8.16)$$

where $k > 0$ is sufficiently small and $a_{ij} > 0$.

Proof: Let $a_{ij} = 0$ if $j \notin \mathcal{N}_i$. Then v_i in (8.16) can be rewritten as

$$v_i = k \sum_{j \neq i} a_{ij} \Big(x_{ij} + R_1(-\phi_i)(c_j - c_i) \Big)$$

$$= k \sum_{j \neq i} a_{ij} \Big((z_j - z_i) + R(\vartheta)(c_j - c_i) \Big) \cdot r_i.$$

After substituting v_i into the kinematic equation of each unicycle, we get

$$\dot{z}_i = kM(\theta_i(t)) \sum_{j \neq i} a_{ij} \Big((z_j - z_i) + R(\vartheta)(c_j - c_i) \Big).$$

Hence, the overall system is given by

$$\dot{z} = kH(\theta(t)) \Big((A \otimes I_2)z + (I_n \otimes R(\vartheta))(A \otimes I_2)c \Big).$$

where $H(\theta(t))$ and A are the same as in (8.10).

By the property of Kronecker product, we know that

$$(I_n \otimes R(\vartheta))(A \otimes I_2) = A \otimes R(\vartheta) = (A \otimes I_2)(I_n \otimes R(\vartheta)).$$

Thus, using the equality above we obtain

$$\dot{z} = kH(\theta(t))(A \otimes I_2) \Big(z - \big(I_n \otimes R(\vartheta) \big) c \Big).$$

Applying the coordinate transformation $\varsigma = z - \big(I_n \otimes R(\vartheta) \big) c$ gives the following new system

$$\dot{\varsigma} = kH(\theta(t))(A \otimes I_2)\varsigma.$$

It follows from the proof of Theorem 8.4 that

$$\lim_{t \to \infty} \varsigma(t) = \mathbf{1} \otimes \xi \text{ for some } \xi \in \mathbb{R}^2.$$

Hence,

$$\lim_{t \to \infty} z_i(t) = R(\vartheta)c_i + \xi, \quad i = 1, \dots, n$$

which means that the n vehicles achieve a pattern formation specified by $c = (c_1, \dots, c_n)$. ∎

Notice that $\theta_i(t) = \theta_i(t_0) + \sin(t)$, so if the n vehicles achieve an agreement on their initial orientation $\theta_i(t_0)$ and treat it as their common sense of direction, the feedback controller (8.16) becomes

$$\begin{cases} v_i = k \sum_{j \in \mathcal{N}_i} a'_{ij} \left(x_{ij} + \mathrm{Re}\left(e^{-j\sin(t)} \right)(c_j - c_i) \right) \\ \omega_i = \cos(t) \end{cases} \qquad i = 1, \dots, n. \qquad (8.17)$$

The agreement on their orientation can be implemented by an linear alignment strategy. However, the controller (8.17) is not robust in practice.

We make a simulation of ten unicycles moving in the plane. Suppose the information flow digraph is the one given in Fig. 8.10. As we mentioned before, it is quasi-strongly connected. A desired circle formation specified by c is considered, where the components of c are

$$c_i = 75 e^{\left(j \frac{2(i-1)\pi}{10}\right)}, i = 1, \dots, 10,$$

respectively. Then, the controller (8.17) is used with the parameter $k = 1$. For the initial condition

$$x(0) = (10, 10, 10, 0, 0, 0, 0, -10, -10, -10)^T,$$
$$y(0) = (-5, 0, 5, 10, 3, -3, -10, -5, 0, 5)^T,$$
$$\theta(0) = \left(\frac{\pi}{6}, \frac{\pi}{6}, \frac{\pi}{6}, \frac{\pi}{6}, \frac{\pi}{6}, \frac{\pi}{6}, \frac{\pi}{6}, \frac{\pi}{6}, \frac{\pi}{6} \right)^T,$$

the trajectories of ten vehicles forming a circle formation are depicted in Fig. 8.21.

8.7 Notes and Discussion

The problem of coordinated control of a group of autonomous wheeled vehicles is of recent interest in control and robotics, see for example, [10, 12, 40, 41, 45, 68, 100, 119, 134, 135]. Over the past decade and a half, many researchers have worked on formation control problems with differences regarding the types of agent dynamics, the varieties of the control strategies, and the types of tasks demanded. In 1990, Sugihara and Suzuki [129] proposed a simple algorithm for a group of point-mass type robots to form approximations to circles and simple polygons. And in the years following, distributed algorithms were presented in [4, 5, 131] with the objective of getting a group of such robots to congregate at a common location. Moving synchronously in discrete-time steps, the robots iteratively observe neighbors within some visibility range and follow simple rules to update their positions. Moreover, there have been many results on the mathematical analysis of formation control. In [85], stability of asynchronous swarms

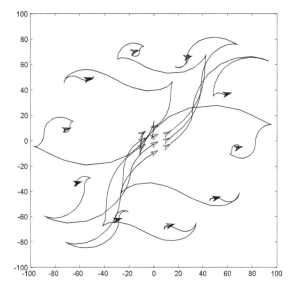

Fig. 8.21. Circle formation.

with a fixed communication topology is studied, where stability is used to characterize the cohesiveness of a swarm. In [115–117], formation stabilization of a group of agents with linear dynamics (double-integrator) is studied using structural potential functions. An alternative is to use artificial potential functions and virtual leaders as in [73]. The approach known as leader-following has been widely used in maintaining a desired formation while moving, e.g., [34, 36, 37].

In addition to the work mentioned so far where a network of vehicles are of linear dynamics, there have been a number of interesting developments in solving the formation control problem for coupled nonholonomic systems. For example, [64] studies achievable equilibrium formations of unicycles each moving at unit speed and subject to steering control and presents stabilizing control laws, wherein each unicycle senses all others (using graph theory terminology, the sensing topology is fully connected), and in [89–91], a circular formation is achieved for a group of unicycles using the strategy of cyclic pursuit, which is a particular form of coupling structure. Considering the facts that no memoryless continuous time-invariant state feedback control law can stabilize a nonholonomic system to the origin as suggested by Brockett [27] and that many efforts were made to find time-varying stabilizing control laws [92, 120, 121] for nonholonomic systems later on, the authors of [147, 148] study the problem of forming group formations of Hilare-type mobile robots by using time-varying feedback control laws. Collision avoidance is another important issue in the coordinated control of a group of autonomous wheeled vehicles. There are some works in the literature based on other formulations which take collision avoidance into account, e.g., [63–66, 149].

We end this section by looking at pseudo-linearization of unicycles.

Consider the unicycle using Frenet-Serret frame

$$\dot{z} = rv$$
$$\dot{r} = s\omega$$
$$\dot{s} = -r\omega.$$

Let $l > 0$. The point $p := z + rl$ is a distance l in front of the unicycle. We then

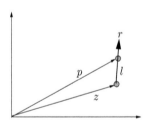

Fig. 8.22. Pseudo-linearization.

have

$$\dot{p} = \dot{z} + \dot{r}r = rv + s\omega l.$$

Define a new control input $u := rv + s\omega l$. Since r, s are orthonormal, the equation

$$rv + s\omega l = u$$

can be solved for v, ω, namely,

$$v = u^T r, \quad \omega = (1/l)u^T s.$$

Then the dynamics of the point p simply

$$\dot{p} = u.$$

Hence, for a collective of unicycle-type robots, if we consider the just-ahead points p_i and the pseudo-control $u_i := r_i v_i + s_i \omega l$ for each robot, we are able to apply the control strategies developed for point-mass robots to unicycle-type robots. However, there are limitations of the approach. First, it is not locally implementable by onboard position sensor. Let us consider the cyclic pursuit strategy for example. That is,

$$u_i = p_{i+1} - p_i.$$

Thus

$$r_i v_i + s_i \omega_i l = z_{i+1} + l r_{i+1} - (z_i + l r_i).$$

We then obtain

$$v_i = (z_{i+1} - z_i)^T r_i + l(r_{i+1}^T r_i) - l$$

and

$$\omega_i = (1/l)(z_{i+1} - z_i)^T s_i + (r_{i+1}^T s_i).$$

The signals $r_{i+1}^T r_i$ and $r_{i+1}^T s_i$, i.e., the components of r_{i+1} in the basis $\{r_i, s_i\}$, cannot be measured by a position sensor on unicycle i. Secondly, when l is very small, the term $(1/l)(z_{i+1} - z_i)^T s_i$ in ω_i would be very large, saturating a real actuator.

Part V

Supplementary Material

A

Stability and Semistability

Consider an autonomous system

$$\dot{x} = f(x) \qquad\qquad (A.1)$$

where $f : \mathbb{R}^n \to \mathbb{R}^n$ is continuous in x on $\mathbb{R}^n$. Let $x(t, x^0)$ be a solution of (A.1) corresponding to the initial condition $x(0, x^0) = x^0$.

A set $\Omega \subset \mathbb{R}^n$ is called a *positively invariant set* of the system (A.1) if for all $x^0 \in \Omega$

$$x(t, x^0) \in \Omega \quad \text{for all } t \geq 0.$$

In the same fashion, a set $\Omega \subset \mathbb{R}^n$ is called an *invariant set* of the system (A.1) if for all $x^0 \in \Omega$

$$x(t, x^0) \in \Omega \quad \text{for all } t \in \mathbb{R}.$$

Invariant sets can be represented by isolated (stationary) points, curves, surfaces and other closed and open sets.

Consider the case where Ω is a set of equilibria of the system (A.1). An *equilibrium* $\bar{x}$ means

$$f(\bar{x}) = 0.$$

Clearly, Ω is an invariant set. Now we introduce the notions of stability and semistability of an equilibrium in Ω [19, 21, 50]. Let $\mathcal{B}_c(x)$ denote the ball in $\mathbb{R}^n$ centered at x with radius c.

Definition A.1. *An equilibrium $\bar{x} \in \Omega$ is called*

- stable *if for all $\varepsilon > 0$ there exists $\delta > 0$ such that*

$$\forall x^0 \in \mathcal{B}_\delta(\bar{x}), \ \forall t \geq 0, \ \|x(t) - \bar{x}\| \leq \varepsilon;$$

- asymptotically stable *if there exists $c > 0$ such that*

$$\forall x^0 \in \mathcal{B}_c(\bar{x}), \ \forall t \geq 0, \ \lim_{t \to \infty} x(t) = \bar{x};$$

- semistable *if there exists $c > 0$ such that*

$$\forall x^0 \in \mathcal{B}_c(\bar{x}), \ \exists \bar{x}' \in \Omega \ stable, \ \forall t \geq 0, \ \lim_{t \to \infty} x(t) = \bar{x}'.$$

According to the definition above, it is easy to say that any equilibrium is asymptotically stable if and only if it is an isolated equilibrium and is semistable. The concept of semistability is used for systems having a continuum of equilibria.

B

Dini Derivative

The investigation of stability analysis of nonlinear systems using an auxiliary function method has been widely studied. However, it may happen that the "natural" auxiliary function is not that smooth. Hence, it is of interest to generalize it to encompass the case of less smooth functions. One of these generalizations is using Dini derivative.

Let a, b $(a < b)$ be two real numbers and consider a function $h : (a, b) \to \mathbb{R}$, $t \mapsto h(t)$ and a point $t^* \in (a, b)$. The *upper Dini derivative of h at t^** is defined as

$$D^+ h(t^*) = \limsup_{\tau \to 0^+} \frac{h(t^* + \tau) - h(t^*)}{\tau}.$$

The notion of lim sup for real-valued functions is defined as follows: for a function $g : \mathcal{D} \to \mathbb{R}$

$$\limsup_{s \to s^*} g(s) = \lim_{\epsilon \to 0} \sup\{g(s) : s \in \mathcal{D} \cap \mathcal{B}(s^*, \epsilon)\}$$

where $\mathcal{D}$ is a domain and $\mathcal{B}(s^*, \epsilon)$ denotes the ball of radius ϵ about s^*. Note that as ϵ shrinks, the supremum of the function over the ball is monotone decreasing, so we have

$$\limsup_{s \to s^*} g(s) = \inf_{\epsilon > 0} \{\sup\{g(s) : s \in \mathcal{D} \cap \mathcal{B}(s^*, \epsilon)\}\}.$$

Lemma B.1 ([114], page 347). *Suppose h is continuous on (a, b). Then h is non-increasing on (a, b) if and only if $D^+ h(t) \leq 0$ for every $t \in (a, b)$.*

In stability analysis, we are interested in the Dini derivative of a function along the solution of a differential equation. Consider the nonautonomous system

$$\dot{x} = f(t, x) \tag{B.1}$$

and let $x(t)$ be a solution. Further, let $V(t,x) : \mathbb{R} \times \mathbb{R}^n \to \mathbb{R}$ be a continuous function, satisfying a local Lipschitz condition for x, uniformly with respect to t.

Thus, the upper Dini derivative of $V(t, x(t))$ with respect to t is given by

$$D^+V(t, x(t)) = \limsup_{\tau \to 0^+} \frac{V(t+\tau, x(t+\tau)) - V(t, x(t))}{\tau}.$$

On the other hand, define

$$D_f^+V(t, x) = \limsup_{\tau \to 0^+} \frac{V(t+\tau, x + \tau f(t,x)) - V(t, x)}{\tau}.$$

The function D_f^+V is called the *upper Dini derivative of V along the trajectory of (B.1)*. It was shown by Yoshizawa in 1966 (see [114]) that

$$D^+V(t^*, x(t^*)) = D_f^+V(t^*, x^*)$$

when putting $x(t^*) = x^*$.

Lemma B.2. *Let $V_i(t,x) : \mathbb{R} \times \mathbb{R}^n \to \mathbb{R}$ be class C^1 for each $i \in \mathcal{I}_0 = \{1, 2, \ldots, n\}$ and let $V(t, x) = \max\limits_{i \in \mathcal{I}_0} V_i(t, x)$. Then*

$$D^+V(t, x(t)) = \max_{i \in \mathcal{I}(t)} \dot{V}_i(t, x(t)),$$

where

$$\mathcal{I}(t) := \left\{ i \in \mathcal{I}_0 : V_i(t, x(t)) = V(t, x(t)) \right\}$$

is the set of indices where the maximum is reached at t.

The proof can be obtained from Danskin's Theorem [29, 32]. Here we present an alternative proof.

Proof: Define

$$h(s) = \min_{i \in I(t)} V_i(s, x(s)) - \max_{i \notin I(t)} V_i(s, x(s)).$$

By the definition of the set $I(t)$, we have $h(t) > 0$. Furthermore, since the function $h(\cdot)$ is continuous, there is an $\bar{\epsilon} > 0$ sufficiently small such that $h(t+\tau) > 0$ for all $\tau \in [0, \bar{\epsilon}]$. This in turn implies that for all $\tau \in [0, \bar{\epsilon}]$,

$$\max_{i \in I_0} V_i(t+\tau, x(t+\tau)) = \max_{i \in I(t)} V_i(t+\tau, x(t+\tau)).$$

Thus, it follows that

$$
\begin{aligned}
D^+V(t, x(t)) &= \limsup_{\tau \to 0^+} \frac{V(t+\tau, x(t+\tau)) - V(t, x(t))}{\tau} \\
&= \lim_{\epsilon \to 0} \sup_{\tau \in \mathcal{B}(0,\epsilon) \cap \mathbb{R}^+} \frac{V(t+\tau, x(t+\tau)) - V(t, x(t))}{\tau} \\
&= \lim_{\epsilon \to 0} \sup_{\tau \in \mathcal{B}(0,\epsilon) \cap [0,\bar{\epsilon}]} \frac{V(t+\tau, x(t+\tau)) - V(t, x(t))}{\tau} \\
&= \lim_{\epsilon \to 0} \sup_{\tau \in \mathcal{B}(0,\epsilon) \cap [0,\bar{\epsilon}]} \frac{\max_{i \in I_0} V_i(t+\tau, x(t+\tau)) - \max_{i \in I_0} V_i(t, x(t))}{\tau} \\
&= \lim_{\epsilon \to 0} \sup_{\tau \in \mathcal{B}(0,\epsilon) \cap [0,\bar{\epsilon}]} \frac{\max_{i \in I(t)} V_i(t+\tau, x(t+\tau)) - \max_{i \in I(t)} V_i(t, x(t))}{\tau} \\
&= \lim_{\epsilon \to 0} \sup_{\tau \in \mathcal{B}(0,\epsilon) \cap [0,\bar{\epsilon}]} \max_{i \in I(t)} \frac{V_i(t+\tau, x(t+\tau)) - V_i(t, x(t))}{\tau} \\
&= \max_{i \in I(t)} \lim_{\epsilon \to 0} \sup_{\tau \in \mathcal{B}(0,\epsilon) \cap [0,\bar{\epsilon}]} \frac{V_i(t+\tau, x(t+\tau)) - V_i(t, x(t))}{\tau} \\
&= \max_{i \in I(t)} \dot{V}_i(t, x(t)).
\end{aligned}
$$

■

From the last equality in the proof, we can see that if V_i is not differentiable at t, then the following is true:

$$
D^+V(t, x(t)) = \max_{i \in \mathcal{I}(t)} D^+V_i(t, x(t)).
$$

C

LaSalle's Invariance Principle

The following is a brief introduction to LaSalle's invariance principle [71, 114].

For the autonomous system

$$\dot{x} = f(x), \qquad\qquad\qquad (C.1)$$

we assume only that $f : \mathcal{D} \longrightarrow \mathbb{R}^n$ is continuous, where $\mathcal{D}$ is an open subset of $\mathbb{R}^n$. With only continuity, uniqueness of solutions is not assured. Let x^0 be a point of $\mathcal{D}$. The initial time is always chosen equal to 0. A non-continuable solution with $x(0) = x^0$ is written as $x : (\alpha, \omega) \to \mathbb{R}^n$, where $\alpha < 0 < \omega$.

The positive limit set of a solution $x(t)$ is designated by $\Lambda^+(x^0)$. A fundamental property of limit sets is stated in the following lemma.

Lemma C.1 ([114], page 364). *If a solution $x(t)$ is bounded, then $\Lambda^+(x^0)$ is nonempty, compact, and connected. Moreover,*

$$x(t) \to \Lambda^+(x^0) \ \text{ as } t \to \omega \quad \text{and} \quad \omega = \infty.$$

We are now ready to state the celebrated theorem—LaSalle's invariance principle.

Theorem C.1 ([114], LaSalle, 1968). *Let $x(t)$ be a solution of (C.1) and let $V : \mathcal{D} \to \mathbb{R}$ be a locally Lipschitz function such that $D^+V(x(t)) \le 0$ on $[0, \omega)$. Then $\Lambda^+(x^0) \cap \mathcal{D}$ is contained in the union of all solutions that remain in $\mathcal{Z} = \{x \in \mathcal{D} : D^+V(x) = 0\}$ on their maximal intervals of definition.*

D

Averaging Theory

Averaging theory is useful for (asymptotical) stability analysis of slowly time-varying systems. More details can be found in $[2, 69, 92, 105, 122]$. The averaging method applies to a system of the form

$$\dot{x} = \epsilon f(t, x, \epsilon) \tag{D.1}$$

where ϵ is a small positive parameter and $f(t, x, \epsilon)$ is T-periodic in t for some $T > 0$.

Associate with (D.1) an autonomous averaged system

$$\dot{x} = f_{av}(x), \tag{D.2}$$

where

$$f_{av}(x) = \frac{1}{T} \int_0^T f(\tau, x, 0) d\tau.$$

Then we arrive at the following theorem.

Theorem D.1 ([69], page 333). *Let $f(t, x, \epsilon)$ be continuous and bounded, and have continuous and bounded derivatives up to the second order with respect to (x, ϵ) for $(t, x, \epsilon) \in [0, \infty) \times \mathcal{D} \times [0, \epsilon_0]$, where $\mathcal{D}$ is a neighborhood of the origin. Suppose f is T-periodic in t for some $T > 0$ and that $f(t, 0, \epsilon) = 0$ for all $t \geq 0$ and $\epsilon \in [0, \epsilon_0]$. If the origin is an exponentially stable equilibrium point of the averaged system (D.2), then there is a positive constant $\epsilon^* \leq \epsilon_0$ such that for all $0 < \epsilon < \epsilon^*$, the origin is an exponentially stable equilibrium point of (D.1).*

E

Berge's Maximum Theorem

Let $\mathcal{X}$ and $\mathcal{Y}$ be subsets of $\mathbb{R}^m$ and $\mathbb{R}^n$, respectively. A *set-valued map* or *correspondence* F from $\mathcal{X}$ to $\mathcal{Y}$ is a map that associates with each element $x \in \mathcal{X}$ a (nonempty) subset $F(x) \subset \mathcal{Y}$.

In the single-valued case, there are two equivalent definitions of a continuous map f at x: the "$\varepsilon - \delta$" definition and the fact that f maps every sequence x_n converging to x to a sequence $f(x_n)$ converging to $f(x)$. Unfortunately, the natural generalizations of these statements to set-valued maps are no longer equivalent.

A correspondence $F : \mathcal{X} \to \mathcal{Y}$ is said to be *upper hemi continuous* at $x \in \mathcal{X}$ if for all open sets $\mathcal{V}$ such that $F(x) \subset \mathcal{V}$, there exists an open set $\mathcal{U}$ containing x such that $x' \in \mathcal{U} \cap \mathcal{X}$ implies $F(x') \subset \mathcal{V}$. A correspondence $F : \mathcal{X} \to \mathcal{Y}$ is said to be *lower hemi continuous* at $x \in \mathcal{X}$ if for any $y \in F(x)$ and for any sequence of elements $x_n \in \mathcal{X}$ converging to x, there exists a sequence of elements $y_n \in F(x_n)$ converging to y.

If a correspondence is both upper hemi continuous and lower hemi continuous, it is said to be continuous. A continuous function is in all cases both upper and lower hemi continuous. Indeed, there are set-valued maps which enjoy one property without satisfying the other. We refer to [8, 9] for more discussions of continuity properties of set-valued maps.

Example E.1. The correspondence F_1 defined by

$$F_1(x) := \begin{cases} [-1, 1] & \text{if } x \neq 0 \\ 0 & \text{if } x = 0 \end{cases}$$

is lower hemi continuous at zero but not upper hemi continuous at zero.

The correspondence F_2 defined by

$$F_2(x) := \begin{cases} 0 & \text{if } x \neq 0 \\ [-1,1] & \text{if } x = 0 \end{cases}$$

is upper hemi continuous at zero but not lower hemi continuous at zero.

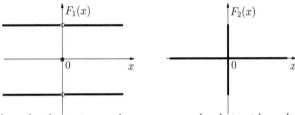

lower h.c. but not upper h.c.   upper h.c. but not lower h.c.

Fig. E.1. Continuity properties of set-valued maps.

Now we are ready to state the Berge's Maximum Theorem.

Theorem E.1 ([130], Berge's Maximum Theorem). *Let* $f : \mathcal{X} \times \mathcal{Y} \rightarrow \mathbb{R}$ *be a continuous function and* $D : \mathcal{X} \rightarrow \mathcal{Y}$ *be a nonempty, compact-valued, continuous correspondence. Then we have*

1. $f^* : \mathcal{X} \rightarrow \mathbb{R}$ *with*

$$f^*(x) := \max \left\{ f(x,y) : y \in D(x) \right\}$$

is a continuous function;

2. $D^* : \mathcal{X} \rightarrow \mathcal{Y}$ *with*

$$D^*(x) := \arg\max \left\{ f(x,y) : y \in D(x) \right\} = \left\{ y \in D(x) : f(x,y) = f^*(x) \right\}$$

is a compact-valued, upper hemi continuous correspondence.

If $D^*(x)$ is a single-valued map, then it is clearly a continuous function.

F

Notation

$\mathbb{R}$, $\mathbb{R}^n$, $\mathbb{R}^{m \times n}$	Real numbers, real n-vectors, real $m \times n$ matrices.
$\mathbb{R}_+$	Nonnegative real numbers.
$\mathbb{C}$	Complex numbers.
$A \succeq 0$, $A \succ 0$	Nonnegative (positive) matrix.
$A \succeq B$, $A \succ B$	$A - B$ is nonnegative (positive) matrix.
A^k	The kth power of matrix A.
$\rho(A)$	The spectra radius of matrix A.
$\ker(A)$	The kernel of matrix A.
$\mathrm{diag}\{a_1, \ldots, a_n\}$	The diagonal matrix with diagonal entries a_i.
$A \otimes B$	Kronecker product of matrices A and B.
$E \sim E'$	Positive and zero entries are in the same position.
$x \succeq 0$, $x \succ 0$	Nonnegative (positive) vector.
$\mathrm{span}\{x_1, x_2\}$	The spanned space of vectors x_1 and x_2.
$\mathrm{Re}[z], \mathrm{Im}[z]$	Real and imaginary part of complex number z.
$\mathrm{g.c.d}\{m, n\}$	The greatest common divisor of integers m and n.
$\mathbf{1}$	The vector of all 1s.
I_m, $I_{m \times m}$	The $m \times m$ identity matrix.
$\mathrm{co}\{p_1, \ldots, p_n\}$	The convex hull of points $p_1, \ldots, p_n$.
$\mathrm{int}(\mathcal{S}), \partial(\mathcal{S})$	The interior and boundary of set $\mathcal{S}$.
$\mathrm{ri}(\mathcal{S}), \mathrm{rb}(\mathcal{S})$	The relative interior and relative boundary of set $\mathcal{S}$.
$\mathrm{lin}(\mathcal{S})$	The carrier space of set $\mathcal{S}$.
$\bigotimes_{i=1}^{n} \mathcal{S}_i$	The Cartesian product of sets $\mathcal{S}_i$.
$\sup_{x \in \mathcal{S}} f(x), \inf_{x \in \mathcal{S}} f(x)$	The supremum and infimum of a function $f(x)$ over $\mathcal{S}$.

References

1. D. Aeyels and P. D. Leenheer, "Extension of the Perron-Frobenius theorem to homogeneous systems," *SIAM Journal on Control and Optimization*, vol. 41, no. 2, pp. 563–582, 2002.

2. D. Aeyels and J. Peuteman, "On exponential stability of nonlinear time-varying differential equations," *Automatica*, vol. 35, no. 6, pp. 1091–1100, 1999.

3. B. D. O. Anderson, "New developments in the theory of positive systems," in *Systems and Control in the Twenty-First Century*, C. I. Byrnes, B. N. Datta, D. S. Gilliam, and C. F. Martin, Eds. Birkhäuser, 1997, pp. 17–36.

4. H. Ando, Y. Oasa, I. Suzuki, and M. Yamashita, "Distributed memoryless point convergence algorithm for mobile robots with limited visibility," *IEEE Transactions on Robotics and Automation*, vol. 15, no. 5, pp. 818–828, 1999.

5. H. Ando, I. Suzuki, and M. Yamashita, "Formation and agreement problems for synchronous mobile robots with limited visibility," in *Proceedings of IEEE International Symposium on Intelligent Control*, Monterey, CA, USA, 1995, pp. 453–460.

6. D. Angeli and P. A. Bliman, "Stability of leaderless discrete-time multi-agent systems," *Mathematics of Control, Signals, and Systems*, vol. 18, pp. 293–322, 2006.

7. M. Arcak, "Passivity as a design tool for group coordination," *IEEE Transactions on Automatic Control*, vol. 52, no. 8, pp. 1380–1390, 2007.

8. J. P. Aubin, *Viability Theory*. Birkhäuser, 1991.

9. J. P. Aubin and A. Cellina, *Differential Inclusions, Set-valued Maps and Viability Theory*. Springer-Verlag, 1984.

10. T. Balch and R. C. Arkin, "Behavior-based formation control for multirobot teams," *IEEE Transactions on Robotics and Automation*, vol. 14, no. 6, pp. 926–939, 1998.

11. J. Bang-Jensen and G. Gutin, *Digraphs: Theory, Algorithms and Applications*. Springer-Verlag, 2002.

12. R. W. Beard, J. Lawton, and F. Y. Hadaegh, "A coordination architecture for spacecraft formation control," *IEEE Transactions on Control Systems Technology*, vol. 9, no. 6, pp. 777–790, 2001.

13. R. W. Beard and V. Stepanyan, "Information consensus in distributed multiple vehicle coordinated control," in *Proceedings of the 42nd IEEE Conference on Decision and Control*, Maui, Hawaii, USA, 2003, pp. 2029–2034.

14. L. W. Beineke and R. J. Wilson, *Graph Connections: Relationships Between Graph Theory and Other Areas of Mathematics*. Clarendon Press, 1997.

15. I. V. Belykh, V. N. Belykh, and M. Hasler, "Blinking model and synchronization in small-world networks with a time-varying coupling," *Physica D*, vol. 195, no. 1-2, pp. 188–206, 2004.

16. ——, "Connection graph stability method for synchronized coupled chaotic systems," *Physica D*, vol. 195, no. 1-2, pp. 159–187, 2004.

17. C. Berge and A. Ghouila-Houri, *Programming, Games and Transportation Networks*. John Wiley and Sons, 1965.

18. A. Berman and R. J. Plemmons, *Nonnegative Matrices in the Mathematical Sciences*. Classics in Appl. Math. 9, SIAM, 1994.

19. D. S. Bernstein and S. P. Bhat, "Lyapunov stability, semistability, and asymptotic stability of matrix second-order systems," in *Proceedings of the American Control Conference*, 1994, pp. 2355–2359.

20. S. Bespamyatnikh, B. Bhattacharya, D. Kirkpatrick, and M. Segal, "Mobile facility location," in *Fourth Int. ACM Workshop on Discrete Algorithms and Methods for Mobile Computing and Communication*, 2000, pp. 46–53.

21. S. P. Bhat and D. S. Bernstein, "Nontangency-based Lyapunov tests for convergence and stability in systems having a continuum of equilibria," *SIAM Journal on Control and Optimization*, vol. 42, no. 5, pp. 1745–1775, 2003.

22. R. Bhatia, *Matrix Analysis*. Springer-Verlag, 1996.

23. F. Blanchini, "Set invariance in control," *Automatica*, vol. 35, no. 11, pp. 1747–1767, 1999.

24. V. Borkar and P. Varaiya, "Asymptotic agreement in distributed estimation," *IEEE Transactions on Automatic Control*, vol. 27, pp. 650–655, 1982.

25. G. Bouligand, *Introduction a la geometrie infinitesimmale directe.* Gauthiers-Villars, 1932.

26. C. M. Breder, "Equations descriptive of fish schools and other animal aggregations," *Ecology*, vol. 35, no. 3, pp. 361–370, 1954.

27. R. W. Brockett, "Asymptotic stability and feedback stabilization," in *Differential Geometric Control Theory: Proceedings of the Conference Held at Michigan Technological University*, R. W. Brockett, R. S. Millman, and H. J. Sussmann, Eds. Birkhäuser, 1983, pp. 181–191.

28. M. Cieliebak and G. Prencipe, "Gathering autonomous mobile robots," in *Proceedings of 9th International Colloquium on Structural Information and Communication Complexity*, 2002, pp. 57–72.

29. F. H. Clarke, "Generalized gradients and applications," *Transactions of the American Mathematical Society*, vol. 205, no. 4, pp. 247–262, 1975.

30. J. Cortes, S. Martinez, and F. Bullo, "Robust rendezvous for mobile autonomous agents via proximity graphs in arbitrary dimensions," *IEEE Transactions on Automatic Control*, vol. 51, no. 8, pp. 1289–1298, 2006.

31. J. Cortes, S. Martinez, T. Karatas, and F. Bullo, "Coverage control for mobile sensing networks," *IEEE Transactions on Robotics and Automation*, vol. 20, no. 2, pp. 243–255, 2004.

32. J. M. Danskin, "The theory of max-min, with applications," *SIAM Journal on Applied Mathematics*, vol. 14, no. 4, pp. 641–664, 1966.

33. M. H. DeGroot, "Reach a consensus," *Journal of the American Statistical Association*, vol. 69, no. 345, pp. 118–121, 1974.

34. J. P. Desai, J. P. Ostrowski, and V. Kumar, "Modelling and control of formation of nonholonomic mobile robots," *IEEE Transactions on Robotics and Automation*, vol. 17, no. 6, pp. 905–908, 2001.

35. J. Desai, J. Ostrowski, and V. Kumar, "Controlling formations of multiple mobile robots," in *Proceedings of IEEE International Conference on Robotics and Automation*, Leuven, Belgium, 1998, pp. 2864–2869.

36. M. Egerstedt and X. Hu, "Formation constrained multi-agent control," *IEEE Transactions on Robotics and Automation*, vol. 17, no. 6, pp. 947–951, 2001.

37. ——, "A hybrid control approach to action coordination for mobile robots," *Automatica*, vol. 38, no. 1, pp. 125–130, 2002.

38. T. Eren, P. N. Belhumeur, B. Anderson, and S. A. Morse, "A framework for maintaining formations based on rigidity," in *Proceedings of the 15th IFAC World Congress*, Barcelona, Spain, 2002.

39. L. Farina and S. Rinaldi, *Positive Linear Systems: Theory and Applications*. Wiley-Interscience, 2000.

40. J. A. Fax and R. M. Murray, "Graph Laplacians and vehicle formation stabilization," in *Proceedings of the 15th IFAC World Congress*, Barcelona, Spain, 2002.

41. ——, "Information flow and cooperative control of vehicle formations," in *Proceedings of the 15th IFAC World Congress*, Barcelona, Spain, 2002.

42. R. Fierro, A. Das, V. Kumar, and J. Ostrowski, "Hybrid control of formations of robots," in *Proceedings of IEEE International Conference on Robotics and Automation*, Seoul, Korea, 2001, pp. 157–162.

43. P. Flocchini, G. Prencipe, N. Santoro, and P. Widmeyer, "Gathering of asynchronous robots with limited visibility," *Theoretical Computer Science*, vol. 337, pp. 147–168, 2005.

44. L. R. Foulds, *Graph Theory Applications*. Springer-Verlag, 1992.

45. J. Fredslund and M. J. Mataric, "A general algorithm for robot formation using local sensing and minimal communication," *IEEE Transactions on Robotics and Automation*, vol. 18, no. 5, pp. 837–846, 2002.

46. V. Gazi and K. M. Passino, "Stability analysis of swarms," *IEEE Transactions on Automatic Control*, vol. 48, no. 4, pp. 692–697, 2003.

47. ——, "Stability analysis of social foraging swarms," *IEEE Transactions on Systems, Man, and Cybernetics–Part B: Cybernetics*, vol. 34, no. 1, pp. 539–557, 2004.

48. N. Gordon, I. A. Wagner, and A. M. Bruckstein, "Gathering multiple robotic agents with limited sensing capabilities," *Lecture Notes in Computer Science*, vol. 3172, pp. 142–153, 2004.

49. J. Gunawardena, "Chemical reaction network theory for in-silico biologists," Bauer Center for Genomics Research, Harvard University, Cambridge, MA, USA, Lecture Notes, 2003.

50. W. M. Haddad and V. S. Chellaboina, "Stability and dissipativity theory for nonnegative dynamical systems: a unified analysis framework for biological and physiological systems," *Nonlinear Analysis: Real World Applications*, vol. 6, pp. 35–65, 2005.

51. J. Hajnal, "Weak ergodicity in non-homogeneous Markov chains," *Proc. Cambridge Philos. Soc.*, vol. 54, pp. 233–246, 1958.

52. F. Han, K. W. T. Yamada, K. Kiguchi, and K. Izumi, "Construction of an omnidirectional mobile robot platform based on active dual-wheel caster mechanisms and development of a control simulator," *Journal of Intelligent and Robotic Systems*, vol. 29, pp. 257–275, 2000.

53. Y. Hatano and M. Mesbahi, "Agreement over random networks," in *Proceedings of the 43rd IEEE Conference on Decision and Control*, Atlantis, Paradise Island, Bahamas, 2004, pp. 2010–2015.

54. D. Hershkowitz and N. Mashal, "P^α-matrices and Lyapunov scalar stability," *The Electronic Journal of Linear Algebra*, vol. 4, pp. 39–47, 1998.

55. J. P. Hespanha, "Uniform stability of switched linear systems: extensions of LaSalle's invariance principle," *IEEE Transactions on Automatic Control*, vol. 49, no. 4, pp. 470–482, 2004.

56. C. Huygens, *Horoloqium Oscilatorium*.   Paris, France, 1673.

57. H. Ishii and B. A. Francis, "Stabilizing a linear system by switching control with dwell time," *IEEE Transactions on Automatic Control*, vol. 47, no. 12, pp. 1962–1973, 2002.

58. A. Jadbabaie, J. Lin, and A. S. Morse, "Coordination of groups of mobile autonomous agents using nearest neighbor rules," *IEEE Transactions on Automatic Control*, vol. 48, no. 6, pp. 988–1001, 2003.

59. A. Jadbabaie, N. Motee, and M. Barahona, "On the stability of the Kuramoto model of coupled nonlinear oscillators," in *Proceedings of the 2004 American Control Conference*, Boston, USA, 2004, pp. 4296–4301.

60. F. Jadot, *Dynamics and Robust Nonlinear PI Control of Stirred Tank Reactors*.   Ph.D. Dissertation, Université Catholique de Louvain, Louvain, Belgium, 1996.

61. K. H. Johansson and A. Speranzon, "Graph Laplacians and vehicle formation stabilization," in *Proceedings of the 16th IFAC World Congress*, Prague, Czech Republic, 2005.

62. V. Jurdjevic, *Geometric Control Theory*.   Cambridge University Press, 1997.

63. E. W. Justh and P. S. Krishnaprasad, "A simple control law for UAV formation flying," University of Maryland, Tech. Rep., 2003, Technical Report No: TR2002-38.

64. ——, "Steering laws and continuum models for planar formations," in *Proceedings of the 42nd IEEE Conference on Decision and Control*, Maui, Hawaii, USA, 2003, pp. 3609–3614.

65. ——, "Equilibria and steering laws for planar formations," *Systems and Control Letters*, vol. 52, pp. 25–38, 2004.

66. ——, "Natural frames and interacting particles in three dimensions," in *Proceedings of the 44th IEEE Conference on Decision and Control*, Seville, Spain, 2005, pp. 2841–2846.

67. T. Kaczorek, *Positive 1D and 2D Systems*.   Springer-Verlag, 2002.

68. W. Kang, N. Xi, and A. Sparks, "Theory and applications of formation control in a perceptive referenced frame," in *Proceedings of the 39th IEEE Conference on Decision and Control*, Sydney, Australia, 2000, pp. 352–357.

69. H. K. Khalil, *Nonlinear Systems*, 2nd ed.   Prentice Hall, 1996.

70. Y. Kuramoto, *Chemical Oscillations, Waves, and Turbulence*.   Springer-Verlag, 1984.

71. J. P. LaSalle, "Stability theory for ordinary differential equations," *Journal of Differential Equations*, vol. 4, pp. 57–65, 1968.

72. P. D. Leenheer and D. Aeyels, "Stability properties of equilibria of classes of cooperative systems," *IEEE Transactions on Automatic Control*, vol. 46, no. 12, pp. 1996–2001, 2001.

73. N. E. Leonard and E. Fiorelli, "Virtual leaders, artificial potentials and coordinated control of groups," in *Proceedings of the 40th IEEE Conference on Decision and Control*, Orlando, Florida, USA, 2001, pp. 2968–2973.

74. D. Liberzon and A. S. Morse, "Basic problems in stability and design of switched systems," *IEEE Control Systems Magazine*, vol. 19, no. 5, pp. 59–70, 1999.

75. J. Lin, A. S. Morse, and B. D. O. Anderson, "The multi-agent rendezvous problem - part 1 the synchronous case," *SIAM Journal on Control and Optimization*, vol. 46, no. 6, pp. 2096–2119, 2007.

76. ——, "The multi-agent rendezvous problem - part 2 the asynchronous case," *SIAM Journal on Control and Optimization*, vol. 46, no. 6, pp. 2120–2147, 2007.

77. Z. Lin, M. Broucke, and B. Francis, "Local control strategies for groups of mobile autonomous agents," *IEEE Transactions on Automatic Control*, vol. 49, no. 4, pp. 622–629, 2004.

78. Z. Lin, M. Broucke, and B. A. Francis, "Local control strategies for groups of mobile autonomous agents," in *Proceedings of the 42nd IEEE Conference on Decision and Control*, Maui, Hawaii, USA, 2003.

79. Z. Lin, B. Francis, , and M. Maggiore, "State agreement for continuous-time coupled nonlinear systems," *SIAM Journal on Control and Optimization*, vol. 46, no. 1, pp. 288–307, 2007.

80. Z. Lin, B. Francis, and M. Maggiore, "Feasibility for formation stabilization of multiple unicycles," in *Proceedings of the IEEE Conference on Decision and Control*, Atlantis, Paradise Island, Bahamas, 2004, pp. 1796–1801.

81. ——, "Coupled dynamic systems: from structure towards state agreement," in *Proceedings of 44th IEEE Conference on Decision and Control, and European Control Conference*, 2005, pp. 3303–3308.

82. ——, "Necessary and sufficient graphical conditions for formation control of unicycles," *IEEE Transactions on Automatic Control*, vol. 50, no. 1, pp. 121–127, 2005.

83. ——, "On the state agreement problem for multiple nonlinear dynamical systems," in *Proceedings of 16th IFAC World Congress*, Prague, Czech Republic, July 4-8 2005.

84. ——, "Getting mobile autonomous robots to rendezvous," *Lecture Notes in Control and Information Sciences*, vol. 329, pp. 119–137, 2006.

85. Y. Liu, K. M. Passino, and M. M. Polycarpou, "Stability analysis of m-dimensional asynchronous swarms with a fixed communication topology," *IEEE Transactions on Automatic Control*, vol. 48, no. 1, pp. 76–95, 2003.

86. D. G. Luenberger, *Introduction to Dynamic Systems: Theory, Models, and Applications*.   John Wiley and Sons, 1979.

87. N. A. Lynch, *Distributed Algorithms*. Morgan Kaufmann Publishers, Inc., 1997.

88. J. Marshall, Z. Lin, M. Broucke, and B. Francis, "Pursuit strategies for autonomous agents," *Lecture Notes in Control and Information Science*, vol. 309, pp. 137–151, 2004.

89. J. A. Marshall, *Cooperative Autonomy: Pursuit Formations of Multivehicle Systems*. Ph.D. Dissertation, University of Toronto, Toronto, Canada, 2005.

90. J. A. Marshall, M. E. Broucke, and B. A. Francis, "Formations of vehicles in cyclic pursuit," *IEEE Transactions on Automatic Control*, vol. 49, no. 11, pp. 1963–1974, 2004.

91. ——, "Pursuit formations of unicycles," *Automatica*, vol. 42, no. 1, pp. 3–12, 2006.

92. R. T. M'Closkey and R. M. Murray, "Nonholonomic systems and exponential convergence: some analysis tools," in *Proceedings of the 32th IEEE Conference on Decision and Control*, San Antonlo, Texas, USA, 1993, pp. 943–948.

93. M. Mesbahi and F. Y. Hadegh, "Formation flying of multiple spacecraft via graphs, matrix inequalities, and switching," *AIAA Journal of Guidance, Control, and Dynamics*, vol. 24, no. 2, pp. 369–377, 2000.

94. L. Moreau, "Leaderless coordination via bidirectional and unidirectional time-dependent communication," in *Proceedings of the 42nd IEEE Conference on Decision and Control*, Maui, Hawaii, USA, 2003, pp. 3070–3073.

95. ——, "Stability of continuous-time distributed consensus algorithm," in *Proceedings of the 43rd IEEE Conference on Decision and Control*, Atlantis, Paradise Island, Bahamas, 2004, pp. 3998–4003.

96. ——, "Stability of multiagent systems with time-dependent communication links," *IEEE Transactions on Automatic Control*, vol. 50, no. 2, pp. 169–182, 2005.

97. R. Murray, Z. Li, and S. Sastry, *A Mathematical Introduction to Robotic Manipulation*. CRC Press, 1994.

98. H. Nijmeijer and A. Rodriguez-Angeles, *Synchronization of Mechanical Systems*. World Scientific, 2003.

99. Y. Oasa, I. Suzuki, and M. Yamashita, "A robust distributed convergence algorithm for autonomous mobile robots," in *Proceedings of IEEE International Conference on Systems, Man, and Cybernetics*, Orlando, FL, USA, 1997, pp. 287–292.

100. P. Ogren, M. Egerstedt, and X. Hu, "A control Lyapunov function approach to multi-agent coordination," *IEEE Transactions on Robotics and Automation*, vol. 18, no. 5, pp. 847–851, 2002.

101. R. Olfati-Saber, J. A. Fax, and R. M. Murray, "Consensus and cooperation in networked multi-agent systems," *Proceedings of The IEEE*, vol. 95, no. 1, pp. 215–233, 2007.

102. R. Olfati-Saber and R. M. Murray, "Consensus protocols for networks of dynamic agents," in *Proceedings of the America Control Conference*, 2003, pp. 951–956.

103. ——, "Consensus problems in networks of agents with switching topology and time-delays," *IEEE Transactions on Automatic Control*, vol. 49, no. 9, pp. 101–115, 2004.

104. J. K. Parrish and W. H. Hammer, *Animal Group in Three Dimensions*. Cambridge University Press, 1997.

105. J. Peuteman and D. Aeyels, "Averaging results and the study of uniform asymptotic stability of homogeneous differential equations that are not fast time-varying," *SIAM Journal on Control and Optimization*, vol. 37, no. 4, pp. 997–1010, 1999.

106. A. Pogromsky, G. Santoboni, and H. Nijmeijer, "Partial synchronization: from symmetry towards stability," *Physica D*, vol. 172, no. 1, pp. 65–87, 2002.

107. W. Ren and R. W. Beard, "Consensus seeking in multi-agent systems under dynamically changing interaction topology," *IEEE Transactions on Automatic Control*, vol. 50, no. 5, pp. 655–661, 2005.

108. W. Ren, R. W. Beard, and E. M. Atkins, "A survey of consensus problems in multi-agent coordination," in *Proceedings of 2005 American Control Conference*, Portland, OR, USA, 2005, pp. 1859–1864.

109. W. Ren, R. W. Beard, and T. W. McLain, "Coordination variables and consensus building in multiple vehicle systems," in *Cooperative Control: A Post-Workshop Volume 2003 Block Island Workshop on Cooperative Con-*

trol, V. Kumar, N. Leonard, and A. S. Morse, Eds. Springer-Verlag, 2004, pp. 171–188.

110. C. Reynolds, "Boids." [Online]. Available: www.red3d.com/cwr/boids/

111. ——, "Flocks, birds, and schools: a distributed behavioral model," *Computer Graphics*, vol. 21, pp. 25–34, 1987.

112. T. Richardson, "Stable polygons of cyclic pursuit," *Annals of Mathematics and Artificial Intelligence*, vol. 31, pp. 147–172, 2001.

113. T. Rockafeller, *Convex Analysis*. Princeton University Press, 1970.

114. N. Rouche, P. Habets, and M. Laloy, *Stability Theory by Liapunov's Direct Method*. Springer-Verlag, 1975.

115. R. O. Saber and R. M. Murray, "Distributed cooperative control of multiple vehicles formations using structural potential functions," in *Proceedings of the 15th IFAC World Congress*, Barcelona, Spain, 2002.

116. ——, "Distributed structural stabilization and tracking for formation of dynamic multi-agents," in *Proceedings of the 41st IEEE Conference on Decision and Control*, Las Vegas, Nevada, USA, 2002, pp. 209–215.

117. ——, "Graph rigidity and distributed formation stabilization of multi-vehicle systems," in *Proceedings of the 41st IEEE Conference on Decision and Control*, Las Vegas, Nevada, USA, 2002, pp. 2965–2971.

118. ——, "Agreement problems in networks with directed graphs and switching topology," in *Proceedings of the 42nd IEEE Conference on Decision and Control*, Maui, Hawaii, USA, 2003, pp. 4126–4132.

119. ——, "Flocking with obstacles avoidance: cooperation with limited communication in mobile networks," in *Proceedings of the 42nd IEEE Conference on Decision and Control*, Maui, Hawaii, USA, 2003, pp. 2022–2028.

120. C. Samson, "Control of chained systems application to path following and time-varying point-stabilization of mobile robots," *IEEE Transactions on Automatic Control*, vol. 40, no. 1, pp. 64–77, 1995.

121. C. Samson and K. Ait-Abderrahim, "Feedback stabilization of a nonholonomic wheeled mobile robot," in *Proceedings of IEEE/RSJ International Workshop on Intelligent Robot and Systems IROS'91*, Osaka, Japan, 1991, pp. 1242–1247.

122. J. A. Sanders, *Averaging Methods in Nonlinear Systems*. Springer-Verlag, 1985.

123. A. V. Savkin, "Coordinated collective motion of groups of autonomous mobile robots: analysis of Vicsek's model," *IEEE Transactions on Automatic Control*, vol. 49, no. 6, pp. 981–983, 2004.

124. R. Sepulchre, D. Paley, and N. Leonard, "Collective motion and oscillator synchronization," in *Cooperative Control: A Post-Workshop Volume 2003 Block Island Workshop on Cooperative Control*, V. Kumar, N. Leonard, and A. S. Morse, Eds. Springer-Verlag, 2004, pp. 89–205.

125. G. V. Smirnov, *Introduction to the Theory of Differential Inclusions*. American Mathematical Society, 2001.

126. I. Stewart, M. Golubitsky, and M. Pivato, "Symmetry groupoids and patterns of synchrony in coupled cell networks," *SIAM Journal on Applied Dynamic Systems*, vol. 2, no. 4, pp. 609–646, 2003.

127. S. H. Strogatz, "From Kuramoto to Crawford: exploring the onset of synchronization in populations of coupled oscillators," *Physica D*, vol. 143, pp. 1–20, 2000.

128. ——, "Exploring complex networks," *Nature*, vol. 410, no. 8, pp. 268–276, 2001.

129. K. Sugihara and I. Suzuki, "Distributed motion coordination of multiple mobile robots," in *Proceedings of the 5th IEEE International Symposium on Intelligent Control*, Philadelphia, PA, USA, 1990, pp. 138–143.

130. R. K. Sundaram, *A First Course in Optimization Theory*. Cambridge University Press, 1996.

131. I. Suzuki and M. Yamashita, "Distributed anonymous mobile robots: formation of geometric patterns," *SIAM Journal on Computing*, vol. 28, no. 4, pp. 1347–1363, 1999.

132. P. Tabuada, G. J. Pappas, and P. Lima, "Feasible formation of multi-agent systems," in *Proceedings of the American Control Conference*, Arlington, VA, USA, 2001, pp. 56–61.

133. ——, "Motion feasibility of multi-agent formations," *IEEE Transactions on Robotics*, vol. 21, no. 3, pp. 387–392, 2005.

134. H. G. Tanner, A. Jadbabaie, and G. J. Pappas, "Stable flocking of mobile agents, part i: fixed topology," in *Proceedings of the 42nd IEEE Conference on Decision and Control*, Maui, Hawaii, USA, 2003, pp. 2010–2015.

135. ——, "Stable flocking of mobile agents, part ii: dynamic topology," in *Proceedings of the 42nd IEEE Conference on Decision and Control*, Maui, Hawaii, USA, 2003, pp. 2016–2021.

136. J. N. Tsitsiklis and M. Athens, "Convergence and asymptotic agreement in distributed decision problems," *IEEE Transactions on Automatic Control*, vol. 29, no. 8, pp. 690–696, 1984.

137. T. Vicsek, A. Czirok, E. Ben-Jacob, I. Cohen, and O. Shochet, "Novel type of phase transition in a system of self-driven particles," *Physical Review Letters*, vol. 75, no. 6, pp. 1226–1229, 1995.

138. V. I. Vorotnikov, "On the coordinate synchronization problem for dynamical systems," *Differential Equations*, vol. 40, no. 1, pp. 14–22, 2004.

139. I. A. Wagner and A. M. Bruckstein, "Row straightening via local interactions," *Circuits, Systems, and Signal Processing*, vol. 16, no. 3, pp. 287–305, 1997.

140. P. K. C. Wang, "Navigation strategies for multiple autonomous mobile robots moving in formation," in *Proceedings of IEEE/RSJ International Workshop on Intelligent Robots and Systems*, Tsukuba, Japan, 1989, pp. 486–493.

141. ——, "Navigation strategies for multiple autonomous mobile robots moving in formation," *Journal of Robotic Systems*, vol. 8, no. 2, pp. 177–195, 1991.

142. J. Wolfowitz, "Products of indecomposable, aperiodic, stochastic matrices," *Proceedings of the American Mathematical Society*, vol. 14, no. 5, pp. 733–737, 1963.

143. O. Wolkenhauer, *Systems Biology: Dynamic Pathway Modeling.* Ph.D. Dissertation, University of Rostock, Rostock, Germany, 2004.

144. C. W. Wu, "Synchronization in arrays of coupled nonlinear systems: passivity, circle criterion, and observer design," *IEEE Transactions on Circuits and Systems–I: Fundamental Theory and Applications*, vol. 48, no. 10, pp. 1257–1261, 2001.

145. ——, "Synchronization in coupled arrays of chaotic oscillators with nonreciprocal coupling," *IEEE Transactions on Circuits and Systems–I: Fundamental Theory and Applications*, vol. 50, no. 2, pp. 294–297, 2003.

146. C. W. Wu and L. O. Chua, "Synchronization in an array of linearly coupled dynamical systems," *IEEE Transactions on Circuits and Systems–I: Fundamental Theory and Applications*, vol. 42, no. 8, pp. 430–447, 1995.

147. H. Yamaguchi, "A distributed motion coordination strategy for multiple nonholonomic mobile robots in cooperative hunting operations," *Robotics and Autonomous Systems*, vol. 43, no. 4, pp. 257–282, 2003.

148. H. Yamaguchi and J. W. Burdick, "Time-varying feedback control for nonholonomic mobile robots forming group formations," in *Proceedings of the 37th IEEE Conference on Decision and Control*, Tampa, Florida, USA, 1998, pp. 4156–4163.

149. F. Zhang, E. W. Justh, and P. S. Krishnaprasad, "Boundary following using gyroscopic control," in *Proceedings of the 43rd IEEE Conference on Decision and Control*, Atlantis, Paradise Island, Bahamas, 2004, pp. 5204–5209.

Index

www.ingramcontent.com/pod-product-compliance
Lightning Source LLC
LaVergne TN
LVHW022304060326
832902LV00020B/3273